Complete Guide to

Preventive and Predictive Maintenance

Joel Levitt

Industrial Press

Library of Congress Cataloging-in-Publication Data
Levitt, Joel, 1952-
The complete guide to preventive and predictive maintenance/by Joel Levitt.
p. cm.
ISBN 0-8311-3154-3
1. Plant maintenance. 2. Industrial equipment--Maintenance and repair. I. Title.

TS192 .L4667 2002
658.2'02--dc21

2002032725

Industrial Press Inc.
200 Madison Avenue
New York, NY 10016-4078

Cover and Text Designer: Janet Romano

First Edition

Complete Guide to Preventive and Predictive Maintenance

4 5 6 7 8 9 10

Disclaimer: The author and the publisher take no responsibility for the completeness or accuracy of any task list contained herein. The lists included are examples only. It is your responsibility to insure completeness. If you want to use these lists or elements of these lists, you must add the proper safety, personnel protection, and environmental protection steps for your particular equipment and operating environment. It is your responsibility to evaluate the individual risks of your equipment, facilities, and environment, and add or change tasks accordingly.

Dedication

This work is dedicated to my personal maintenance consultant Hall of Fame. These people taught me much of what I know about maintenance consulting. They also represent integrity, quality and good value for their clients.

Semond Levitt, my father was the prototypical consultant interested about any topic that came into his attention. Jay Butler was my first consultant employer and had his own unique beliefs about fleet maintenance, many of which I now share. Don Nyman is a colleague, collaborator on a book on planning and trained me when I started in the field. Ed Feldman trained and advised me on custodial maintenance. Ricky Smith and Richard Jamison generously let me work with their consultants on larger projects and gave me insight into larger maintenance consultancies. Mark Goldstein has been an infatiguable friend, a mentor and teacher with a unique insight into the best role for maintenance. Mike Brown for his low-key approach to many maintenance fields.

Another group affected me because of their presence in the field. Some are my friends too. One of the leaders of the field is John Campbell whose gentle guidance has impacted the direction of the whole field. John Moubray for his excellent work in RCM had single handedly changed my views about the role of PM. Terry Wireman for his written contributions to the field and endless speeches. From down under Sandy Dunn has trained a generation of maintenance thinkers.

List of Figures

Table of Contents

Usage of Terms

In this book the words asset, unit, equipment and machine are used interchangeably. In some industries they have different meanings. In this book, all these words mean the basic unit, system, or machine that the PM is addressed to. To confuse the issue, some CMMS use their own special terms to refer to same concepts or items.

PM means Preventive Maintenance
PdM means Predictive Maintenance
PPM or **P/PM means** Preventive and Predictive Maintenance
CMMS means Computerized Maintenance Management System

PM for this book does not mean:

Pencil Maintenance (where the inspector skips the inspection and pencil whips the form.

Panic Maintenance (actually our PM is just the opposide)

Planned Maintenance (Its related, like a first cousin but not even a sibling)

Productive Maintenance (Our PM can help you get there.)

Or finally

Percussive Maintenance

(The fine art of whacking an electronic device to get it to work again.)

Introduction

How to view PM (Preventive Maintenance) and PdM (Predictive Maintenance)

In prior works by this and other authors, PM has been treated as an engineering issue (what tasks will have the greatest impact?) or as a management issue (as in procedures and preparation for TPM). Other writers have considered PM as a combination of ways (RCM- engineering, and economic aspects).

In fact, PM is even more complicated than the above considerations. Effective PM or PdM is like a skyscraper with four sides. PM initiatives commonly fail to meet expectations or just gradually fade out of existence when one side is neglected. If the program is to be successful it needs to have structural integrity in all four areas:

Engineering: The tasks have to be the right tasks, being done with the right techniques, at the right frequency. Many PM systems have elaborate PM tasking but breakdowns occur anyway because the wrong things are being looked at in the wrong frequency. In other words, the tasks have to detect or correct critical wear that is occurring. Analysis of statistics of failure, uptime, and repair is included in the engineering pillar of PM.

Economic: The tasks must be 'worth' doing. One measure of worth is that doing the tasks furthers the business goals of the organization. Is the value of the failure greater than the cost of the tasks? Spending $1000 to maintain an asset worth $500 is usually a waste of resources unless there is a downtime, environmental, or safety issue. This economic question is critical. The RCM approach includes in the 'worth doing' equation, except where tasks where failures could result in environmental catastrophe or loss of life or limb. Many PM initiatives ignore the consequence of failure and are discontinued (properly) because they are not worth the effort.

People-Psychological: The people doing the PM have to be motivated to the extent that they actually do the designated tasks properly. Without motivation, PM rapidly becomes mind numbing. PM people also need to attend to the level of detail generated by a PM system and they must be properly trained to know what they are looking at and why.

Management: PM has to be built into the systems and procedures that control the business and these systems must be designed so that good PM results. W.E Demming, the quality guru, said that quality was in the system of production not in the individual effort. A tacked-on PM system is rarely effective for the long haul. Information collected from PM has to be integrated into the flow of business information. PM data has to be reported to the Plant Manager or Director of Operations so that there is a structure outside maintenance asking questions, demanding answers, and demanding accountability.

This book is designed to address all four aspects of PM

The Holy Grail of Maintenance

PM is the best approach, right? After 5000 years of Preventive Maintenance (they used to do PM inspections on the great pyramids) why is the PM approach not the dominant one? Why do we still need to argue for resources? Why do over 70% of all organizations with physical assets only have a rudimentary PM system or none at all?

The keys to this mystery are in two related areas. One is in human nature. Human nature makes us very reluctant to invest time and resources in something that 'might' happen. We are mostly short-term beings, interested in results now, not a year from now. The PM approach is all about what might happen. If there were any certainty to PM's predictions, then selling PM would be simplified. It is argued that you could spend all your time chasing what-ifs, and never get any real work done.

The second key is in answering the question, what does top management really want from us (the maintenance function)? This is a non-trivial question because many managers cannot verbalize the answer. Maintenance professionals are thus left serving many masters (different daily opinions based on the pressures of the day embodied into one person)

The Holy Grail of maintenance is (paradoxically) to get out of the maintenance business. Your job will become clear when you understand this!

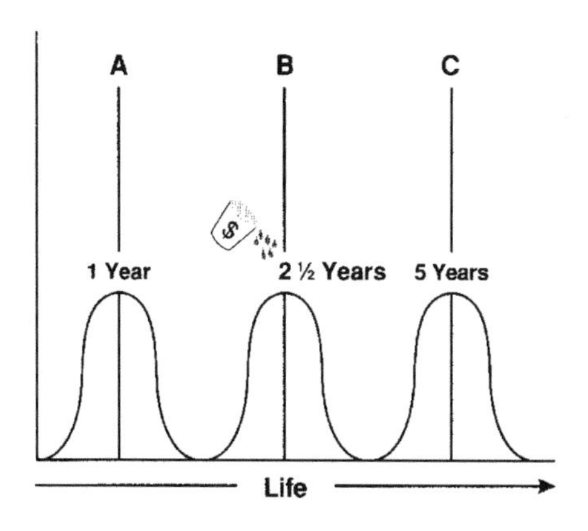

The illustration shows three ways to manage physical assets like your factory, facility or fleet.

The X-Axis is time or utilization. As you travel to the right more time has elapsed, more product has been made, or more mileage has been driven.

The Y-axis is the number of breakdowns or disruptive incidents. The more breakdowns there are, the higher the curve. Eventually everything wears out. The first breakdown in that plant is on the left side of the curve just where the curve becomes visible. By the same token, the last breakdown in that plant is on the right side of the curve, where it hits the X-axis.

Each scenario reflects the average life of all equipment in the facility under that program. For example, if you operate a fleet of cars and don't do any maintenance the cars will have a certain average breakdown rate. If you add PM activity the average life will be extended. In the last curve to the right you re-engineer the cars to be more reliable. Every time you get some breakdowns you look for designs that are more reliable and modify the equipment

Curve Descriptions

1. BS Breakdown Scenario: This curve is closest to the Y-axis so that, on average, more events take place in a shorter interval. The breakdown scenario means that no PM is done, or no effective PM is done. In this environment, chaos reigns. Some days it is really quiet and some days everything is broken (and your most important customer visits!)

Don't knock it; in certain industries this BS might be the best way to run. Look for situations where equipment is low value and can be replaced cheaply and quickly and where there are low production needs and low quality requirements. BS just might be a low cost alternative.

However, the breakdown scenario has consequences. The environment is usually chaotic, and full of high stress, and routinely requires heroism just to get production out the door. The level of safety (incidents per 100,000 hours of operation) might be higher than a PM dominated shop of the same size and type. People tend to burn out. Almost all organizations start here when they are small. Unfortunately most organizations stay with this curve as they react to what happens. The breakdown of machines creates the schedules for both production and maintenance labor.

One other aspect in this scenario is that you are likely to get a lot of BS. There will always be excuses about why this machine didn't run and why that job is not complete.

2. PM scenario- you go out looking for problems. You take specific steps to extend the life of the equipment and to detect impending failure. The focus here is on investigation of the critical wear points so that breakdowns are deferred as long as possible and repairs or replacements are made before failure occurs.

This procedure is also known as proactive maintenance. In this sense, proactive means a maintenance orientation toward tasks performed today that detect or defer future breakdown. Companies (and entire industries) can get very good at this trick and experience fewer and fewer breakdowns. The nuclear power industry comes to mind. A good proactive maintenance process is the goal for many of the top firms.

Equipment lasts longer with proactive maintenance but the PM scenario requires money, thought, and management. Many firms are unwilling to commit the money or management talent to such a goal. The huge problem is that once money stops being poured into PM, the plant or building reverts to the breakdown scenario.

3.MI Maintenance Improvement, (the Holy Grail) of the maintenance department means – get out there and fix the problems permanently. Fix them in such a way that the expected failure rate drops to a tenth or a fifth of what it was. This objective is one of the stated goals of RCM.

Solving problems permanently is one of the most rewarding aspects of maintenance. Ask any maintenance old-timer and you'll have frequently, a long discussion of redesign, re-purposing, re-specification, and re-engineering.

Is it possible to operate without breakdowns? The same old timers who will regale you with stories of successful re-engineering, will tell you "never!" – "Not possible." Yet all of us have equipment that never fails, that in spite of a complete lack of maintenance the pump, compressor, or press, runs and runs. Why not study the reasons for such longevity instead of spending time thinking so much about breakdowns. In other words, let's get out of the repetitive repair business. This vision means the death of maintenance, as we know it.

What happened?

Sometime in the 90's the old way of doing business died and went away. We might mourn the loss of some of the positive aspects of that world but, for better or worse, it is truly gone. The old paradigms and strategies are obsolete in light of the new corporate order. Our corporate sponsors (the same developments occurred in the public sector too) realized that they needed something different from the maintenance function to face new, tougher, no-holds-barred competitors (or lower tolerance for increased taxes in public sector organizations). We must now ask fundamental structural questions about what types of tasks maintenance personnel ought to do and who should do maintenance tasks. The first question of this inquiry is what is the mission of maintenance?

What is the mission of maintenance?

There used to be many different answers to this question (as many as there were organizations asking the question). The mission definitions ranged from quick reaction times in fixing breakdowns to serving the customer more efficiently. Some firms are intent on reducing downtime and others focus on cost control or quality. A few focus on safety or environmental security. All these missions are good, useful, and important. All of them ignore the deep issue that the organization has changed and that there is something very simple that transcends these missions or values.

In today's organizations the creed is that everyone must add value to the product. Everyone and everything is expendable, outsourcable. There is a conflict between the old mission statements and the new culture. The new mission is:

"The mission of the maintenance department is to provide reliable physical assets and excellent support for its customers by reducing and eventually eliminating the need for maintenance services."

New roles

This new mission requires a re-thinking of traditional roles. On one side, maintenance must merge with machine, building, and tool design to integrate maintainability improvements into designs on an ongoing basis. The accumulated knowledge and lessons of maintenance will be merged immediately into the design profession. There will be a revolving door between the people who design and the people who maintain.

On the other side, routine maintenance activity will be merged increasingly into operations. The TPM model shows that the operator is capable of this integration and the whole maintenance effort will benefit from operator involvement.

The consequences of breakdowns must be managed!

There is a traditional attitude on the part of maintenance that all breakdowns are the same and all are equally bad. (After all, if it's broken it's broken). This acceptance of the status quo is now intolerable and unacceptable in maintenance. A breakdown should be viewed with an analytical eye to see what difference it made (if any). Any money spent must be justifiable in light of the consequences of failure. By the way, failures that result in death, serious injury, or environmental damage, are not acceptable at all! Any equipment that requires periodic attention to avoid breakdowns is likewise a failure of design engineering.

Where does PM and predictive maintenance fit into the new structure?

There are two situations where PM (and PdM) is important. One situation is when it reduces the probability or the risk of death, injury, or environmental damage to zero or near zero. The second situation is where the cost of the task is lower than the cost of the consequences of the failure. If this rule sounds familiar, it should, because it has become the mantra of the RCM movement. That rule is the beginning but not the whole conversation.

As addressed in the Holy Grail discussion, the fatal flaw of the old type of PM and PdM is that they require constant investment of labor and materials. In most instances, no relationship is traced between the cost of the consequences of the failure and the cost of the PM service. The financial relationship between failure consequences and tasks must be built into the system from the beginning. PMO (PM Optimization) makes great strides in alignment of the task costs to the failure mode consequences.

There is another problem. PM institutionalizes the status quo. No permanent improvement will ever flow from a traditional PM orientation. When you are downsized and PM is deferred, the MTBF (Mean Time Between Failures) curve will return to its old breakdown frequency. So the second idea is that the third curve, the curve of maintenance improvement, must be added into the priorities of the department. Return to the new mission: "to provide excellent support for its customers by reducing and eventually eliminating the need for maintenance services."

In this context there is a place for PM in the new organization. First and foremost, view PM as a manager of consequence. To eliminate maintenance efforts look at PM as a way station or resting-place on the way to maintenance elimination. When you don't have the time, resources, or technology to figure out the root cause of a failure you can use a PM approach to reduce your exposure to breakdown and its consequences. Of course, you must also continue PMs in addition to other methods where the implications of breakdown are deadly or very expensive.

How is maintenance to be created with the new mission?

Continuous improvement in the delivery of maintenance is the new goal. The bulk of management time, money, and effort must go to reducing the labor, parts, utilities, and overhead or to increasing uptime. The stakes are high. What is at stake could be the survival of your organization. There are competitors who are eyeing your market share and they are not standing still.

Groundwork
PM, PdM Defined For This Book

There are many ways to look at the whole field of PM and PdM. For this book we will use the model shown here, which has been successful because it incorporates all maintenance activity in a structure designed to reduce maintenance activity. The 'Holy Grail" of Chapter 1 is achieved over time within a secure structure of PM (in the diagram it is called design and engineering review).

By building one structure for all activity the policies and procedures can be more easily explained and roles assigned to all maintenance personnel. For success, it is essential that analysis be completed in a timely manner and that all personnel have a role. Highly technical analysis is sometimes best achieved by maintenance craftspeople (with proper training).

A maintenance department run from the model discussed here will have some attributes. Note that little structure is designed to manage breakdown. The focus of this model is keeping equipment running, not in repairing broken equipment. A year of analysis, redesign, and corrective maintenance will move an organization out of the breakdown mode.

PM is a series of tasks performed at a frequency dictated by the passage of time, the amount of production (cases of beer made), machine hours, mileage, or condition (differential pressure across a filter) that either:

1. Extend the life of an asset. Example: Greasing a gearbox will extend its life. All the tasks with 'E' in the box are extend type tasks

-Or-

2. Detect that an asset has had critical wear and is about to fail or break down. Example: A quarterly inspection shows a small leak from a pump seal. Finding this leak allows you to repair it before a catastrophic breakdown. All the tasks with 'D' in the box are detect type tasks

Description of the details from the chart on page 7.

Row 1

Task List: a list of all the tasks or actions to be performed at that time (there are 4 major types of task lists- unit, string, standing, and future benefit).

Types of clocks and frequency: How often or when to perform tasks on the task list. Measured in days, units, tonnage, cycles, miles, or even readings (such as temperature), changes to readings, findings (oil slick on floor under truck). Almost any trigger can be designed into a PM clock.

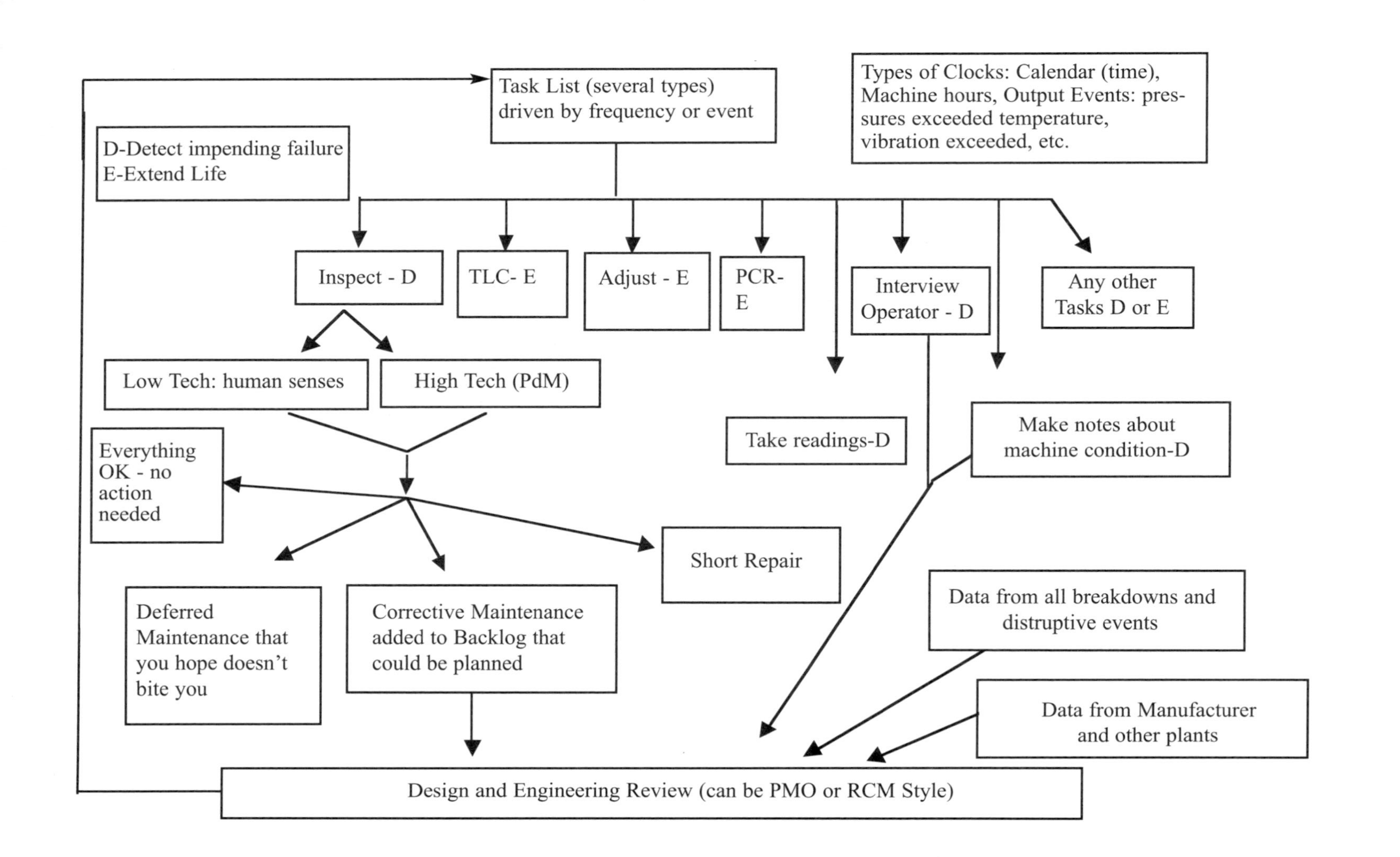

Task List (several types) driven by frequency or event
Types of Clocks: Calendar (time), Machine hours, Output Events: pressures exceeded temperature, vibration exceeded, etc.
D-Detect impending failure
E-Extend Life
Inspect - D
TLC- E
Adjust - E
PCR- E
Interview Operator - D
Any other Tasks D or E
Low Tech: human senses
High Tech (PdM)
Take readings-D
Make notes about machine condition-D
Everything OK - no action needed
Short Repair
Deferred Maintenance that you hope doesn't bite you
Corrective Maintenance added to Backlog that could be planned
Data from all breakdowns and distruptive events
Data from Manufacturer and other plants
Design and Engineering Review (can be PMO or RCM Style)

Row 2 and 3 (D-detect failure, E-extend life, these rows represent possible types of tasks)

Inspect: Stop, look, and listen, using human senses or instruments (PdM) **(also row 3)**

PdM (Predictive maintenance): Any inspection carried out with high technology tools that use advanced technology to detect when failures will occur. Such tools can increase your returns and give you more time to intervene before failure. **(also row 3)**

TLC (tighten, lube, clean): Start with the basics. Caring for your equipment is the core of the PM approach. This care does not require any fancy equipment or techniques, just basic care. Much of the benefit from PM flows from TLC.

Adjust: Making the equipment work optimally by tightening, changing, fine-tuning, or modifying the machine set-up or operation.

PCR (Planned Component Replacement): also called scheduled replacement. One of the tools in your pouch is PCR. This technique has been made popular by the airlines. PCR can improve reliability in many circumstances.

Readings: Writing down or entering data concerning measurements of pressure, temperature or other parameters. Spotting trends in these readings can frequently uncover problems before they impact production or safety.

Interview operator: Ask questions about machine operation and note answers. Many problems are apparent to the operator or driver before they are obvious to anyone else.

Notes about machine condition: These notes are related to readings and will tell the skilled observer of any subtle changes taking place in the asset.

Row 4

There are four outcomes from a PM inspection:

1. Everything okay - no action needed

2. Deferred maintenance item. – You will ignore this problem and hope the unit doesn't fail. The problem is that these deferred items have a way of coming back to haunt you. They only rarely go away by themselves. Deferred maintenance items have been studied and when looked at economically they tend to deteriorate at a great rate compared to the interest rate that could be received on the money not spent by deferring them in the first place. In common language it rarely pays to defer an action unless the machine or process is being closed down.

3. Corrective maintenance: is any item found by inspection that you plan to schedule. We call this plannable maintenance (it can be planned). The goal of the inspection process is plannable maintenance. With this kind of work you have the lead time to work efficiently. We say plannable not planned because not every firm is committed to planning maintenance activity (for details see *Maintenance Planning, Scheduling and Coordination,* Industrial Press)

4. Short repairs: these are repairs done by the PM person when they are doing the PM, including repairs of short duration with the tools and materials that the PM person carries. These actions are different from temporary repair. A short repair is a complete repair that can be accomplished in a short time. This subject is discussed in depth in a later chapter. Short repairs are an easy way to improve productivity.

Row 5

All data flows to the design and engineering review. One of the primary reasons for collecting data is to use it in the review (and redesign) of breakdowns and disruptive events. These events include data from breakdowns, data from manufacturers, readings, reports of machine condition, and all work orders. RCM or PCO style design and engineering review uses the structures of RCM to manage the process.

RCM (Reliability Centered Maintenance: One of the most important approaches to PM, and was developed in the aviation industry. One result from a review of what happened is feedback to the task list in the form of details of increased (decreased) frequency, depth, or technology.

PMO (PM Optimization) is an offshoot of RCM and recognizes the difficulty (and sometimes futility) of RCM in a mature operational plant. PMO embodies techniques to optimize the PMs that are done to get the most reliability from the least resources.

Quick look at current efforts by the stakeholders

Consider calling a meeting of all interested parties to talk about the subject of PM. Perhaps start with the report card and discuss the differences between the scoring (frequently the differences are more telling than the areas of agreement). Use the opportunity to teach the group about one or two aspects of PM.

Later in that meeting, or at another meeting, give out lists of the 10 questions and have people think deeply on the subject of whether the company has been serious about PM in the past and if it is truly committed for the future.

Schedule a few meetings where some of the questions can be kicked around and an essential element of PM can be taught. PM knowledge is not well distributed in the organization. A high level of knowledge about the inner workings of PM is very helpful for long-term success.

Ten Questions to start a PM discussion with your staff and managers

1. Does top management support the PM system with its attention, money, and authorizations for downtime as required?

2. Is involvement in PM activity considered high status among the workers?

3. When deficiencies are found by inspection, are they written up as scheduled work and completed in a reasonable time?

4. Do repeated or expensive failures trigger an investigation to find the root cause and correct it?

5. Was there an economic analysis of each task list proving ROI (Return on Investment)?

6. Is PMO (PM Optimization), Reliability Centered Maintenance (RCM) considered when equipment failure could cause injuries, the equipment is critical or has high downtime costs.

7. Are units outside the PM system because they are in very bad shape and fixing them is not worthwhile?

8. Does the failure history impact the frequency, depth, and items on the task list?

9. Did your staff design or modify the design of the task list?

10. Are PM personnel consulted when designing new processes, machines, or buildings?

Rate your PM effort.

Another way to look at PM is to rate your organization's performance at key PM tasks. This kind of evaluation can precede the ten questions or follow it. Remember, knowing where you are is essential before embarking on any journey.

What do you do well, not so well, and not at all? What should be added to your current task lists?

PM Activity	Grade
Inspection (human senses)	
Hi-tech inspections (predictive maintenance)	
Cleaning program	
Tightening bolts	
Lubrication program	
Checking operation of unit	
Minor adjustments	
Take and trend readings	
Planned replacement, overhaul or planned discard	
Interview operator	
Periodic Analysis	
Excellent record keeping	
Checking unit history	
Failure analysis	
RCM or PMO analysis	
Other:	

Getting in the PM Ballpark. How much PM is enough?

Most books about PM say that there is an optimum amount of PM for every kind of operation. Intuitive thinking about the topic also says there must be an optimum level. Too much PM and you waste the labor for the extra inspections. Too little PM and you don't catch breakdowns. PM impacts other cost areas (we discuss this in the next chapter under PM economics). This curve can get complicated when other areas are considered.

What drives the "kind of operation" are things like the cost of downtime, the amount of government regulation, the value of the materials that go into the product (PM in a turbine blade maker might be high), the involvement of the public, hazards involved, visibility, and overall economics

B- All breakdown costs, including such costs as, quality, etc. These items are termed costs below the waterline in the next chapter.

PM- All PM and PdM costs, including labor and material costs and downtime to accomplish PM

T- All costs of maintenance, downtime, labor, parts, above and below the waterline.

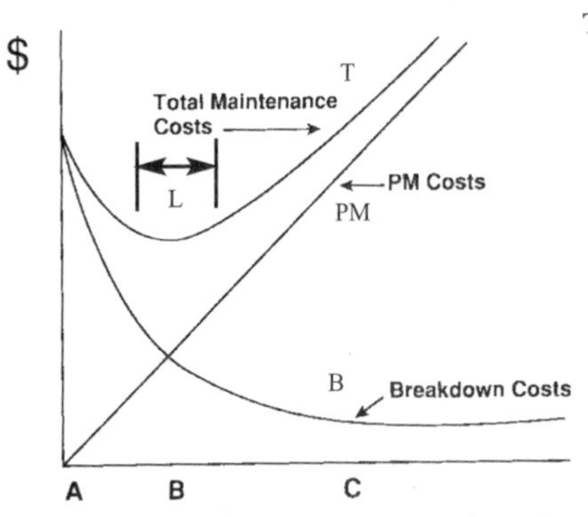

T- Total Maintenance Costs and non-maintenance costs below the waterline.

PM costs: Increasing as tasks are added. Downtime for PM included

L- Area of lowest overall cost

B- Breakdown costs start going down as PM increases but at somepoint the costs start to stabilize as PM gets to be less effective

Curve showing lowest overall costs based on increasing PM costs to an optimum level

L- The goal is to have the lowest overall costs of production or lowest costs to deliver a service. You can see that the lowest overall costs results from a certain level of PM, no more and no less.

There are many variations of this curve that add other layers to arrive at the most accurate representation of the relationship between PM and overall costs. Some authors properly add in the cost of equipment downtime as a separate curve. Others add customer goodwill, quality, and other important cost areas to make the contributors to the total cost highlighted.

In fact the curve starts to approach the total cost of production where there might be tens of additional variables (which will make the curve hard to read). In its simplest form the curve tells an important story. There is a level of PM that is optimum for your operation.

PM as a Percentage of Hours: Where to start to look for your optimum curve?

For starters, in an industrial setting of no particular industry we would look for:

% **Direct maintenance labor**	**Activity**
15-25%	PM and PdM task list activity
55-60%	Corrective Maintenance as a result of the task list inspections plus short repairs
15-30%	Breakdowns and other customer initiated work.

If you are in a particularly dangerous or regulated environment such as airlines, nuclear power, or pharmaceutical goods, the breakdown level had better be less than 2-5%! An apartment building on the other hand can tolerate higher levels of breakdown (particularly of non-critical assets).

History of the PM movement

Lyon Sprague De Camp wrote about some of the oldest PM efforts in his very interesting book *Ancient Engineers*. He traces the works of engineers from the beginning of recorded history through to the Middle Ages. The Egyptians, with their pyramid cities of the dead, certainly needed inspections and remediation to keep everything in good shape. In those days the designers, builders, and maintainers were the same people.

The irrigation projects of the Tigris and Euphrates river valley supplied water to an entire civilization and required constant inspection and repair. Not much was left from any of the early civilizations of the PM task lists, corrective work orders, or inspection schedules, to indicate how maintenance was done.

The Roman public works such as aqueducts, roads, and public buildings were unsurpassed for almost a thousand years. The Romans had significant PM exposure with elimination of leaks from the aqueducts. Without a vigorous inspection program there would be no water for the customers at the end of the run. The aqueducts were used to deliver water to the homes of the powerful senators and other heavyweights, so any lack of water was a significant problem for the engineer in charge (if he wanted to keep his job and his life!).

One of the most interesting examples of a PM approach was the maintenance of the Roman roads. On some of the more highly traveled roads the engineers used lead shapes as keys to hold the paver-stones in place. Inspectors had to travel the roads periodically to replace these keys because the populace would remove them and sell them. Lead was valuable and widely used. These roads allowed Rome to manage a far flung empire and allowed knowledge, goods, and people to move about the ancient world.

The fall of the Roman empire signaled the decay of the public works that the Romans built and maintained. The roads were torn up and the stone was used for local buildings, aqueducts fell into disrepair, and there was not the will or skill to make long-term repairs. Kingdoms were required to maintain the roads in their domain, but few had the extra resources and will to do so. The subsequent decline in travel, trade, and exchange of ideas contributed to the dark ages and the contracting European life.

Indirectly, lack of PM led to the 500-year decline of European civilization. We can see something like this today. When an organization cuts PM and doesn't keep its assets usable, it is on a down hill slide. Ruins are the end of vitality for an area. Unless there is a massive investment the decay accelerates and it becomes harder to rebuild each year (until it is a pile of rubble).

Throughout history, wherever there were large-scale works such as the roads or pyramids, behind the scenes there were (mostly) men keeping them in good shape. If we jump ahead to modern times, PM is not only accepted but actually discussed by engineers and others as ways to support the mission of the organization.

In the modern times the military has given the most attention to PM as a way to improve reliability. In fact, you can read the arguments in a 1919 issue of the magazine for motor pools. An article in that issue was about inspections versus breakdowns for the 40,000 trucks left in Europe after WWI.

Peter S. Kindsvatter gives us some interesting history of PM in the US military in his article titled *PREVENTIVE MAINTENANCE IN WORLD WAR II.* He says

> "Successful commanders have always understood the importance of what, in today's Army, is called "preventive maintenance." During the Revolutionary War, for example, General George Washington chastised his officers for allowing their men's muskets to rust and fall into disrepair. He directed that soldiers who lost their bayonets or allowed their weapons to be damaged through negligence were to have the costs of repair or replacement deducted from their pay. That Washington had to repeat these orders several times during the course of the war indicates not only his awareness of the importance of preventive maintenance, but also the difficulties inherent in enforcing maintenance discipline.

Kindsvatter goes on to describe the enormous level of effort required to get preventive maintenance to become a regular part of the routine. The War department had to completely reorganize the way vehicles were managed to get a handle on maintenance.

After WWII there was gradual recognition that increasingly complex equipment required some organized actions to insure availability. Both the military and the private sector were involved in their early modern discussions.

Any PM effort in the factories just after WWII probably fell by the wayside (as did many of the advanced quality initiatives pioneered by Demming and others) in the push to get product out of the door. Many shop floor veterans then started to realize that PM approaches were important.

By 1952-3 we find the US Navy Bureau of docks hiring Sears to design a PM system to manage their extensive docks and buildings holdings. Sears came up with a laminated card system which was probably a predecessor to the Visifile Card System still sold today.

The computer would eventually change everything. In 1965 Mobil Oil designed an IBM 360 program named MIDEC designed to manage lubrication schedules on forklifts and other mobile equipment. This program was generally accepted as the first (or one of the first) computerized PM systems.

P/PM Economic

Economics of PM has three levels of view. The highest level might be called macro economic analysis. In this view the firm decides whether PM approaches make sense, given the organization's goals and the needs and requirements of the business or field.

To make such a decision an organization would look at the current costs of operation and would project the costs of the operation using the proposed changes. Since any change costs money the analyst would see how many months or years the savings (assuming there is any) would take to pay off the investment.

If speed of pay-off (which is 1 over the Return on Investment –ROI) were adequate then the decision would be made to change from the status quo to the new approach. Once that decision is made, the second level looks more and more closely at groups of machines or processes.

Speed of impact in such an analysis is of the essence. At the time of this writing, companies are under tremendous pressure to increase profits. In today's business climate, a payback of 3 months is commonly required to get people's attention (to at least have neutral impact on the quarterly results). Of course, good investments that take a year to pay off are never undertaken.

For want of a better term I use semi-micro view for the second level. This semi-micro view decides what strategy is most appropriate for a particular machine or group of machines. Even if a decision has been made at the corporate or plant level to use PM/PdM as the dominant strategy, each machine or machine group has factors that influence how to apply it specifically.

Usually the most important factor is the cost of having the unit out of service (downtime cost). A low or negligible downtime cost can scuttle a PM decision for that asset. As described above, the cost of your current operation for that asset or asset group is compared to the cost of running in the new mode. Given the investment level to bring the asset to PM standards, the question becomes "is there enough Return on Investment (ROI) to justify the cost?"

Once a decision is made about strategy for an asset or an asset group the third level asks what PM tasks do we perform?

In the task view or micro view, the cost and consequence of each task is compared with the cost and consequence of the failure mode the task is trying to avoid. It is critical to choose the fewest tasks that will achieve your goals.

Macro Case study: Effects of investment decisions on the bottom line and the future of the company

Tony Cavanaugh is the president of Springfield Manufacturing. He's been in the steel fabrication business since 1989 and has seen a lot of changes. Tony prides himself on the fact that Springfield is a profitable fabricator with a net income of about 7% of sales.

Barbara Strathmore, Springfield's sales manager, has proposed an expansion into some new areas that would require purchasing an automated plasma burner. This expansion would require $100,000 in new equipment. The expansion would bring in an estimated $500,000 in new sales revenue starting the next year and would also cut costs on current jobs (but the amount of the savings could not easily be calculated).

Tom Duvane, maintenance manager has proposed investment in a new computerized PM system. Calculations show the return would come from reduced parts in stock, increased up time and reductions in maintenance costs. Tom also said that the system would allow his existing staff to support more equipment. The investment would be $75,000 with returns of $75,000 in the next and subsequent years (at present utilization figures).

Tony is not inclined to make both investments in the same year. Which one is better, and which one should be done first (even if the returns on both were the same)?

Item	Sales department	Maintenance Department
Return per year	Revenue * net profit %	Savings per Year
ROI: Return per year/ Investment	**35%** ($500,000 * 7%) /100,000	**100%** = $75,000/$75,000
Payback: Investment/ Return per year	**2.87 years** = 100,000/ ($500,000 * 7%)	**1 year** = $75,000/$75,000

Comparison of results from investments

Clearly the maintenance investment is superior from a purely economic point of view. Most business professionals don't realize that maintenance improvement funds flow directly to the bottom line. A maintenance cost reduction is worth 5-25 times more in profit than similar increases in revenue.

With many business issues the economics only captures a small part of the opportunity. Some questions to consider are: Is a window of opportunity being missed by not buying the new machine (such as a good customer who is being forced to look for a fabricator with a plasma cutter)? How is Tom Duvane, maintenance manager, going to compete with the investment opportunities offered by the sales department, and other departments? After all without some "spin" maintenance can be deadly boring.

Macro view of PM

The macro view of maintenance takes the current cost of operation and compares it with the new cost of operation after changes have been made. If the cost is lower we say there will be a Return on Investment (ROI). If the cost is higher we hope to be able to prove our case for overall lower costs on line items outside the traditional maintenance budget.

To determine the maintenance budget, each machine has to be analyzed for each type of maintenance exposure.

1. PM cost development lists every piece of equipment that has PM activity. PM activity can be determined from a close look at the scheduled services (hopefully) in the computer. This list would normally be a spreadsheet with at least these 5 columns.

PM's per year	Hours per PM	PM Hours per year	Parts per PM in $	Parts per year in $ all PM services

2. CM (corrective maintenance) is a result of all the PM inspection activity. How many hours will be spent next year on corrective maintenance (which consists of short and long repairs initiated by a write-up during an inspection)? CM is not directly part of the PM system but is certainly part of the PM program. Without it, the impending failure is detected but not repaired.

Number of Incidents	Hours per year	Material Costs per year

3. UM is all user requests including breakdowns, small projects, special requests, etc: These data can be determined from a close look at history

Number of Incidents	Hours per year	Material Costs per year

4. RM represents demand created by rebuilds, new installations, rehabilitation, and capital projects associated with the asset. Usually RM is a separate line item in the budget. Sometimes it is handled by the maintenance budget when a project runs out of budgeted funds and the project is not complete.

Number of Incidents	Hours per year	Material Costs per year

When all the different demands are put together for an asset or asset group one line of analysis (which can be used as a zero-based budget) has been completed. In the next year there would be comparisons of budget to actual for each asset or asset group. The point of this exercise is to run the numbers in the present condition and run them again with the changes you contemplate. Then you may compare and decide which is the winner.

Much of this data can be gathered from your CMMS. Run the report into a spreadsheet and go on your way. The number of incidents and services will help you plan the size of your support effort, which will approximate to the number of work orders issued.

Asset # or Group	PM Hr	PM Mat'l	CM Hr	CM Mat'l	UM Hr	UM Mat'l	RM Hr	RM Mat'	# Incidents

What happens when we ignore the signs?

What benefit is it to repair a chemical transfer pump before failure?
What benefit is it to repair a roof before failure?
What benefit is it to repair a wheel bearing before failure?
What benefit is it to repair a drill press before failure?

We've spoken about the cost of downtime and the value of being able to schedule when the repair takes place. These are important reasons. There is another reason to make the repair before the failure occurs. It is damage to associated parts. When assets fail they frequently damage associated parts. You can either replace the bearing on the chemical transfer pump or the impellor, housing, and other items. With this collateral damage, the failure might triple in cost.

A roof is an excellent example of this concept in action. A small leak can be fixed for a small amount of money. But left to its own devices the leak will destroy everything in its path including ceilings, inventory, and machinery.

Think about a wheel bearing that is allowed to fail in a vehicle. That failure could not only cause an accident but cause very expensive damage to the axle and other parts. The rule is, the longer you wait the more unpredictable the outcome. What was a well-controlled problem can blossom into a complete nightmare!

Sometimes the failure doesn't get worse and the cost and damage are pretty much the same. A cheap drill press might be discarded so that it is perfectly ok to run it to failure. But be sure you've thought through the consequences before you start down that road.

When you defer action you stick the future maintenance department with an increasingly expansive problem.

One problem in factories, fleets and buildings...

Some PM systems abandoned. What is meant by failure? Most PM systems are rolled out to much fanfare. A visit a year or two later shows the PMs are not being done with regularity. Corrective items are being deferred. Inspector's reports are being shelved. Unplanned events are as high as always and the people on the shop floor have excuses. Lastly, the paperwork is either incomplete or information is being faked. The worst part of PM failure is that management might not know the system failed! It is analogous to having the patient die but the hospital to continue to send bills for new services.

There are a variety of reasons for failure, which may be due to mismanagement, bad economics, misreading psychology, or inadequate engineering. The most common cause of failure is plain and simple economics. In any economic analysis, whether macro- or –micro, there is one type of hidden cost usually responsible for sinking the ship.

PM systems fail because **PAST SINS** wreak havoc on anyone trying to change from a fire fighting operation to a PM operation. Even after running with PM for a few months there are still so many emergencies that it seems you can't make headway.

You face **unfunded maintenance liabilities.** The only way through this jungle is to pay the piper, modernize, and rebuild, your equipment until you are out of the woods. This is where the investment must be made. Any sale of a PM system to top management must include a non-maintenance budget line item for past sins.

Remember the wealth was removed from the assets and equipment when they were used without maintenance funds being invested to keep them in top operating condition.

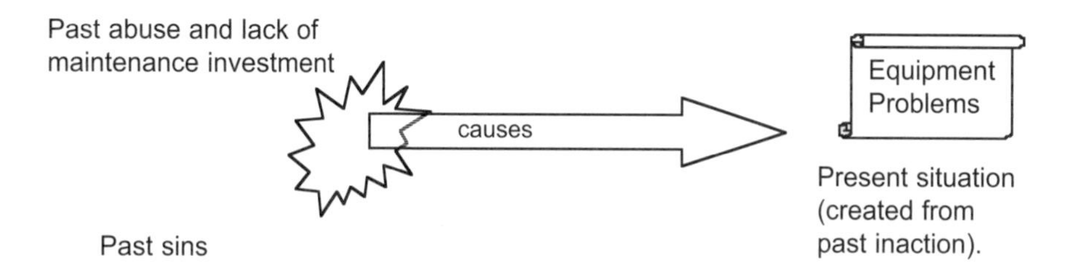

Case Study in pulling back from the brink of disaster

This sequence can be illustrated in a high profile manufacturer that made a very well known product. Quality and profitability had slipped for several years. If it weren't for its core of rabid boosters the company would only be a footnote to the industry. Management engineered a buyout from the conglomerate that had owned it during its period of decline.

Management was faced with an aged physical plant (there were machines still in daily service that had seen their best days before WWII). The company was faced with a product having serious quality issues and a dealer network ready to bolt. The management and the workers labored around the clock and one night they converted the whole plant to a Just-in-Time manufacturing model.

They swept through and processed all the work in process and ran with 8 hours of inventory on the floor. Everything went incredibly well until the first, second and subsequent breakdowns. Here was the rub. Their old equipment was not in good enough shape for it to be relied on it to produce products in a time critical environment. When an eight-week supply of each part was being made, as formerly, a breakdown was not as critical.

Management decided to install a PM immediately and set out to hire 8 new PM inspectors. However, with some analysis the managers realized that the PM inspectors wouldn't help the problem. Past sins were frowning on them and they had to pay the piper, examine the plant (starting with the most critical items), so that the equipment could be put into good solid condition before a PM system could be of any benefit. The recommendation, which was accepted, was to hire 8 new equipment technicians to go through the plant and fix up, bring back, and modernize all the equipment. The technicians also undertook a massive effort to replace the worst equipment.

Costs of a PM system:

One time: Modernization of equipment to PM standard (pays for past sins) including parts, OT, contractors
Costs for training and facilitation for everyone to change the culture
Cost for system (CMMS) to store information
Indirect system costs (such as wiring computers, supplies, extra computer seat licenses, etc)
Data entry labor for data collection (additional labor for auditing data entered)
Labor to train inspectors
Labor to setup task lists and frequencies,
Labor to create job package plans and set standards for all PM routines
Purchase of predictive maintenance devices with training

On-going: Labor for PM task lists, short repairs
Parts costs for task lists, PCR's
Additional investments in predictive maintenance technology
Funds to carry out write-ups (maintain the higher standard of maintenance)
On-going training
Business changes to keep PPM going

Breakdowns

One of the most powerful incentives is to recapture some of the costs below the waterline on the maintenance iceberg, which floats 90% submerged. The costs above the waterline are all on the maintenance manager's budget. They include maintenance labor, parts, supplies, contractors, maintenance overhead, etc. In most companies the costs above the line are a small part of the total cost of even a minor shutdown.

The list below starts to capture the kind of hidden costs, some of which are below the waterline.

You can sell PM by looking deeply at the effect of downtime from breakdowns. In time- critical businesses, reduction of breakdowns and downtime can be strong selling points. The challenge is developing data on the costs.

Cost of Breakdown

What are some of the hidden costs for an average breakdown?

Production Costs

Cost of lost production
Cost of airfreight of finished goods to customer
Spoilage, contamination, or other compromise to product
Loss of goodwill, loss of customers or re-rating from an A vendor to a B vendor
Cost of overhead associated with original assignment
Loss of smooth function

Extra Maintenance Costs

Extra costs due to core damage (destruction of a normally rebuildable part)
Extra damage to associated parts and the labor to repair them
Incoming airfreight
Extra costs of outside vendor parts and labor
Operator (crew) idle time
Extra travel time for mechanic
Extra repair time due to conditions

Accumulate your average number of breakdowns per year and compare 70% of that cost to the cost of the inspections, adjustment, lubrication, and short repairs. We assume that 70% of the breakdowns will be eliminated through an average quality PM system. The formula below should be used in arguments aimed at proceeding with adoption of a PM system.

(Cost of all breakdowns * 70%) > cost of PM system

Case Study #1: Air Compressor

1. Breakdown does not cause safety or environmental exposures.
2. Compressor breaks down every 2 years on average (or 50% probability each year)
3. Repair costs are $15,000 each time
4. Downtime costs are $45,000 per incident

Breakdown costs

(**$30,000**) per year = 0.5<probability of failure in 1 year> * ($45,000<down time costs> + $15,000<repair costs>)

Compare the annual costs under the breakdown plan with the costs of PM
PM costs
PM Service compressor 12 times per year at a cost of $1000 per service (Parts, labor, downtime) plus a biennial scheduled overhaul costing $16,000
New probability of failure with PM: 1 failure in 20 years = 0.05

US $23,000 = (($1000<PM cost> * 12<services per year>)+
(.5<frequency of overhaul>* $16,000<cost of overhaul>)) +
(0.05<new probability of breakdown>($45,000<downtime costs> +
$15,000<repair costs>))

With the above assumptions, PM clearly costs less than the cost of the breakdown mode but it is not enough less to make the decision a forgone conclusion. The cost is greater than 70%, so more analysis is needed. One thing that would help is a redesign of the PM system to lower the $1000/month service cost. Another improvement would be to delay the biennial overhaul to 30 or 36 months if possible.

$30,000 * 70% = $21,000 (allowable)
Current PM cost = $23,000

Case Study 2:
PM Alternatives with cost justifications and consequences

This case study regarding a chemical transfer pump is in the middle level. At the macro level the firm that owns this pump has already decided to use PM strategies where they are appropriate. The analysis now drops to the level of deciding what to do with a particular asset or asset class. After this analysis is complete the firm would proceed to the task level or microanalysis of what specific tasks should be done.

Facts:

Repetitive failure in a chemical transfer pump has been occurring for the last several years. Downtime from lost production is valued by the cost accounting department at $500 an hour after the first hour (no cost for the first hour). Labor hours are valued at $40.

Engineering analysis shows that the application is severe and that this is the best that can be expected. The skilled mechanics working with engineering have designed a PM task list that will drop the number of failures dramatically.

Currently, in breakdown mode, the pump is failing 4 times a year. Each incident requires 10.0 hours and $2000 of parts to get the pump back on line. Due to a reservoir in front of the next process the pump can be out of commission for up to an hour before the downstream processes are affected so that in this particular situation the first hour of downtime is free. Downtime from calling maintenance to full operation is 14 hours (2 hours to respond to the call and 2 hours to get the system filled up and back in operation).

Breakdown costs =	Probability of breakdown in any year	(Cost of breakdown +	Cost of Downtime)
$35,600	**4**	(10 *$40 labor) + $2000 **material & Labor = $2400**	(14-1 hours) * $ 500 **downtime = $6500**

We will use this breakdown cost as the number to beat in all the subsequent scenarios. One hour is deducted from downtime because the first hour is free. A probability of 4 gives 4 breakdowns per year on average. If downtime were not part of the calculation the breakdown costs would drop dramatically.

What are consequences of this choice?

The domain of consequences includes both directly economic (such as replacement parts) and indirectly economic (such as safety) costs. Other consequences include disruption to a smooth operation, customer impact, need to work excessive hours, ongoing disruptions, morale issues, environmental contamination, etc.

Every breakdown is a disruption to the smooth running of the business. Each one is an emergency. If the whole plant is run this way there will be a good deal of overtime for the maintenance staff and a hurried atmosphere. Production staff will be very nervous because their capacity could drop out at a moment's notice.

Ongoing Summary:

Name Strategy	Number of incidents per year	Annual Cost	Labor hours	Downtime hours	Capital Investment	Need for management	Reliability and manageability of operation
Breakdown	**4**	**$35,600**	**40**	**56 or 52***	**None or Low**	**Low**	**Low**

*Chargeable downtime

With any machine there are 10 or more basic strategies for maintenance management (many, many more if you consider combinations and hybrids). Some of these strategies are beyond the scope of this book. Some strategies (that we will not analyze here) might include designing a quick-change pump ready to slide into place whenever there is a breakdown. A quick-change approach is to engineer a replacement pump so that it can be changed quickly. When the pump breaks down, the mechanic can literally slide the replacement pump into place, perhaps using quick-connect fittings for power and intake/output. This approach provides an optimized breakdown methodology.

Another strategy is to mount a back-up pump in the system. This approach is called redundancy. Given adequate space, money, and the high costs of downtime, a back-up pump will give the highest level of reliability (which is one reason why commercial aircraft have two or more engines)

Strategies could also include changes in business approach such as outsourcing pumping capability (You can buy just compressed air and the vendor of the air has to worry about maintenance of the compressor, why not do the same with pumping a fluid?).

The **PM routine** designed by the mechanics will take 1 hour a month, $10 of material and will require downtime (remember the first hour is free) to accomplish. In addition to the above considerations it was estimated that 4 hours of administrative time annually were expended, which we will add into the PM servicing time. Additional corrective maintenance incidents occur 3 times annually and each one takes 5 hours and $2000 worth of materials. With the new PM program, breakdowns will drop to 1 every other year (same cost structure as before: 10 hours + $2000 parts and 13 hours of downtime).

PM	(Cost for PM services)	(Cost of Corrective Maintenance incidents with downtime) +	(New probability of breakdown * (Cost of breakdown + Cost of Downtime)
$720 + $12,600 + $4,450 **Total $21,770**	$600 = (12 * $40 labor) + (12 * $10 materials) $160 = 4 * $40 (adm) **Total PM= $720**	3 times/YR ((5 * $40 labor) + $2,000 materials) +(6 hr * $500 downtime) **Total Corrective Maintenance: $16,600**	**$4,450** = 0.5 ($2,400 + $6,500)

You will note that the breakdown cost was taken directly from the last example but with a new probability of occurrence (0.5 is every two years). The corrective maintenance downtime was reduced by 1 hour because the first hour is free. A modest amount (4 hours) was added to reflect administrative costs for a year. This figure assumes that there are many other analyses going on and each will contribute 3, 4, or more hours to the administrative kitty.

What are consequences of this choice?

The PM choice has several consequences and requires certain types of support. There is no PM without support. The support comes in two forms. PM requires administrative time. This example included 4 hours annually of administrative time. While that might not sound like a lot of time, only a few hundred machines are needed to pay for a PM coordinator. Without that person, the supervisor must do the support him/herself and has less time to supervise.

Unlike the breakdown option, a company must plan for and schedule PM activity. The second kind of support is access to the asset when the PM is scheduled and when a corrective action is needed. If the monthly inspections and the corrective maintenance work orders are not completed in a timely manner the whole scenario will slip back into the breakdown mode.

In this choice, the number of unscheduled events drops from 4 a year to one every other year. At the same time the total number of times the pump is touched dramatically increases from 4 to 15.5 times per year. One thing that is interesting (and common) is the increase in the number of downtime incidents. Scheduled downtime is a frequent consequence of a PM approach. With scheduled downtime there is no downtime cost exposure for short downtime incidents, which is fortunate.

PM requires a certain mindset to be effective. If you're prone to ignore equipment, this scenario will require a major change in thinking. Without that change the plan is more wishful thinking then effective business strategy.

Lets look at a comparison between the two choices so far.

Ongoing Summary:

Name Strategy	Number of incidents per year	Annual Cost	Labor hours	Downtime hours	Capital Investment	Need for management	Reliability and manageability of operation
Break-down	4	$35,600	40	56 or 52*	None or low	Low	Low
PM	**15.5**	**$21,770**	**31**	**19.5* or 31.5**	**None or**	**Moderate**	**High**

* Chargeable downtime

PCR (Planned Component Replacement)

Another PM alternative is Planned Component Replacement or Planned rebuild. This approach is related to the quick-change method except that the pump is changed out on a planned basis before failure.

PCR takes that idea a step further. In this example a PCR interval of 2 months would be required. The PCR operation would take 2 hours of downtime and mechanic time. Bringing the pump back to operational specifications would take 5 hours each time and $500 worth of materials. The new failure rate would be once in 10 years (with similar costs as the breakdown case).

PCR Cost =	(Cost per PCR * # Of replacements per year)	(New probability of breakdown * (Cost of breakdown + Cost of downtime)
$25,320 + $890 = $26,310	$25,320 = 6 ((5 * $40 labor) + (5 * $500 material) (3 * $500 downtime)) + 3hr * $40 administration	$890 = 0.1 ($2,400 + $6,500)

As in previous examples, the original cost of breakdown is used with modification as to how often it occurs. We've dropped the administrative contribution to 3 hours a year because there were fewer incidents to schedule. As before, the first hour of downtime is free. This scenario would be much more attractive ($18,810 - almost as good as the PM scenario) if we could re-engineer the system to be changeable in an hour or less.

What are consequences of this choice?

PCR is an interesting alternative. It is widely used (but most maintenance professionals don't know that is what they are doing when they proactively change belts annually or replace consumption items on a predetermined schedule before failure. This particular strategy results in a very reliable environment of 1 failure in ten years (a probability of 0.1).

Economically this scenario falls between breakdown and PM. It requires some support but less often than PM. The number of incidents is what must be managed. Since the machine is touched so much less often, the amount of support is quite a lot less. In either approach, the absolute amount of support is low but essential for success.

The amount of scheduled disruption in PCR is high and would be unacceptable in some circumstances. In others the bi-monthly change out could easily be accommodated.

Lets look again at a comparison between the three choices.

Ongoing Summary

Name of Strategy	Number of incidents per year	Annual Cost	Labor hours	Downtime hours	Capital Investment	Need for management	Reliability and manageability of operation
Breakdown	4	35,600	40	56 or 52*	None or	Low	Low
PM	15.5	$21,770	31	19.5* or 31.5	None or	Moderate	Moderate
PCR	**6.1**	**$26,310**	**31**	**24 or 15***	**Moderate**	**Moderate**	**High**

* Chargeable downtime

As you can imagine, when you add up every variation and combinations of variations there are literally hundreds of alternatives for managing a group of assets. There is no RIGHT answer. There just are consequences that are desirable for your operation and consequences that are undesirable for your operation.

Selling PPM to Management: Battle for the Share of the MIND

It won't shock anyone to know that maintenance professionals as a whole are not great marketers! Maintenance offers some of the best investments possible, both in terms of cost reduction (above the waterline) and production increases (below the waterline). Yet many maintenance investments are viewed with boredom or out and out distrust.

There is a Vicious Cycle of Maintenance (partially adapted from PM Optimization by Steve Turner) that, without intervention, will destroy management's efforts to improve the quality, reliability, and long-term profitability. The first step for gaining mind share is to teach this concept and then to continually revisit this concept at every opportunity.

Vicious Cycle of Maintenance

The vicious cycle is the beginning of a story that has a happy ending in an effective PM system. That system is in an extremely competitive battle for the organization's investment dollars. Investments in maintenance can earn big returns. The returns come from efficiencies in all areas of the operation.

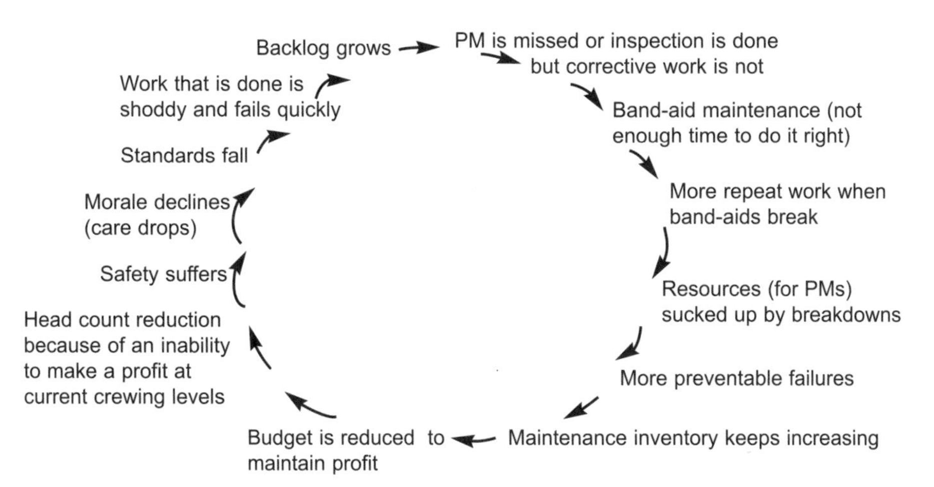

Marketing means to 'Spin' the maintenance story to topics that the audience is vitally interested in

We must sell our strong suits, which are cost avoidance, improved customer satisfaction, and reduced downtime. Use the language (and issues) of your organization to sell a PM program. In every organization there are issues that are more important than any others. You must sell your improved maintenance management investments using these issues.

PM activity has been proven in study after study to lower the cost of operations and improve reliability. In a 1985 article published by ASME called Progress and "Payout of a Machinery Surveillance and Diagnostic Program", the authors Hudachek and Dodd report that rotating equipment maintained under a PM model costs 30% less to maintain versus a reactive model. It further states that adding predictive technologies adds significant additional return on investment.

Real Benefits of a PM system

---

1. Reduces the size and scale of repairs
2. Reduces downtime (increase uptime)
3. Increases accountability for all cash spent
4. Reduces number of repairs
5. Increases equipment's useful life
6. Increases operator, maintenance mechanic and public safety
7. Increases quality of output
8. Reduces overtime for responding to emergency breakdown
9. Increases equipment availability
10. Decreases potential exposure to liability
11. Reduces investment by not needing spare or stand-by units
12. Insures that all parts are used for authorized purposes
13. Increases control over parts, reduce inventory level
14. Decreases unit part cost 15. Improves information available for equipment specification
16. Lowers overall maintenance costs through better use of labor and materials
17. Lowers cost/unit (cost per ton of coal, cost per widget, cost per student) 18. Improves identification of problem areas to show where to focus attention.

Thousands of bits of information arrive during the day. Everyone in your organization filters the information so that they only have to listen to topics that will have an impact on them. How good are you at seeing the world from someone else's point of view? Put yourself in the shoes of top management, plant manager, or accounting. Can you see the world from their concerns, interests, problems and very importantly from the point of view of enhancing their career

Why is this list of PM benefits important?

Why is it important to know the priorities of your listeners? How would this change if you were talking to maintenance workers, bus drivers, or machine operators? Every type of employee or role in your firm has a point of view and interests that they hold dear. When proposing a change to the way business is conducted, consider these outside points of view. The success of your effort depends on what people do when you are not there.

Identify your organization's (and its individuals') priorities: Speak to these priorities

PM Basics

PM is four dimensional

As discussed in the introduction, the four dimensions of the PM or PdM programs are: Economic, Engineering, Psychological, and Management. If we do not consider implications along all four dimensions we will not realize the full benefits possible from these programs. Effective PM must consider impacts and implications in all four dimensions.

Chapter 3 dealt with the first dimension of PM, including a complete discussion of the economics of PM. Without an economic motivation there is no PM. For some practitioners the economic side is the only important one. The PM activity must make economic sense but it is clearly not sufficient that it only make economic sense. On the other hand, many PM task lists have never been vetted by an economic analysis. Tasks are mistakenly being done where the economic consequences of the failure is far less severe than the PM costs.

RCM (Reliability Centered Maintenance) adds depth to the economic discussion as it explicitly adds in the factors of safety, environmental damage, and downtime. Once the impacts of these three additional factors are added to the mix the case for PM becomes several notches more persuasive.

Chapters 4, 6, 10-14, and 17 all deal with different aspects of PM engineering. It is essential that the reason for the failure be well understood. Without that understanding the tasks might not make much sense. Engineering will deliver the right task, to the right component, using the right tool, at the right frequency, to avoid or detect the failure. Engineering also helps us by re-engineering the machine to make PM tasks easier, safer, quicker to complete, to require only one person and to use less material, etc. In some sense, good engineering can help ameliorate problems in task economics (making the tasks cheaper to perform) and in mismatched psychology (making tasks easier and capable of being performed faster).

Many of us have had first hand experience with the mind-numbing repetitiveness of PM tasks. For some people, stringing PM's like lubrication routes, vibration routes are the worst. Why are some people better at PM than others? Why are two, good, talented mechanics so different when it comes to sticking to the task list? The answer is in their psychologies. In Chapter 18 we investigate some of the aspects of psychology that impact both the willingness and the capability to be effective in PM activity. The best PM system will fail without people doing the tasks as designed and at the right frequencies.

The second aspect is personality. What kinds of people make the best PM inspectors? Are there personal attributes that you should look for in your inspectors? Finally, realize that the specifics of your individual PM environment will affect the type of person needed for the job.

Why is it that useless paperwork in organizations never seems to die yet a valuable PM system goes away at the blink of an eye? The answer is management structures. Management structures insure that correct procedures occur whether or not the plan's booster is present. In other words, these management structures keep procedures in place over time.

Consistency is a great challenge. In PM we need structures to continue the inspection and task list completion. Usually the CMMS print a schedule and reports on PM compliance (these are structures to keep the good practices in place). Another aspect is recording the corrective items and completing them in a timely manner. A powerful protocol is needed to insure that the work is done before failure.

In Chapter 15 we discuss the structures necessary to keep PM on the table. All programs for improvement are launched with enthusiasm and fanfare. The single biggest challenge of management is to both preserve and be able to recreate this improved state in the months and years to come.

Are your PM approaches complete (in the view of the four dimensions)? When you are thinking about your own PM program, consider the adequacy of your approach to each of the four dimensions. You may be surprised to find that one or more dimensions might be missing. Be aware that organizations that pride themselves on engineering or economics (etc.) might have that dimension handled to the exclusion of all else.

Six misconceptions about PM

1. PM is only a way of trying to determine when and what will break or wear out so that you can replace it before it does.

> PM is much bigger than that. It is an integrated approach to budgeting and failure analysis, and permanent correction of problem areas. I also eliminates excessive use of resources, and can actually be seen as a way of life!

2. PM systems are all the same. You can just copy the system from the manual or from your old job and it will work.

> PM systems must be designed for the specific equipment as set-up, age of the equipment, product, type of service, hours of operation, skills of operators and many other factors.

3. PM is extra work on top of existing workloads and it costs more money.

> PM increases uptime, reduces energy usage, reduces unplanned events, reduces airfreight bills, etc. There are hundreds of ways PM saves the organ-

ization resources. The only time PM is an addition to the existing workload is at the startup when you put a PM system into place. You will have to spend extra to fund monies not invested in the equipment in the past (pay for past sins).

4. With good forms and descriptions unskilled people can do PM tasks.

   Unskilled (in maintenance) people can do some of the PM tasks successfully with good training and clear forms. For the greatest return on investment, skilled people must be in the loop. TLC activities (such as lubrication, cleaning, or tightening bolts) can certainly be done by almost any trained employee but not maintenance employees. Generally, inspection benefits greatly from experienced eyes and hands.

5. PM is obsolete because of newer technologies

   All proactive maintenance activity is part of PM. That includes the most modern approaches such as vibration routes, infrared surveys, or condition based maintenance checks. The newest PM strategies initiate activity on some condition (such as initiate task list when temperature gauge reads 220°).

6. PM will eliminate breakdown

   In the words of a PM class "PM can't put iron into a machine." In other words the equipment must be able to do the job. PM cannot make a 5-hp motor do the work of a 10-hp motor. Even with the most advanced PM there will still be breakdowns from abuse, misapplication, or accident. Some failure modes (such as electronics failures) do not currently lend themselves to PM approaches.

Task List

The task list is the heart of the PM system. It reminds the PM inspector what to do, what to use, what to look for, how to do it, and when to do it. *In its highest form, the task list represents the accumulated knowledge of the manufacturer, skilled mechanics, and engineers, in the avoidance of failure.*

The best task lists could be designed by a variety of stakeholders including OEMs (original equipment manufacturers), skilled mechanics, engineers, contractors, insurance companies, governmental agencies, trade associations, equipment distributors, consultants, and sometimes by large customers.

All task list items are designed to perform one of two functions. The two functions are the core of all PM thought and they are either extending the life of an asset or detecting when the asset has begun its descent into breakdown (before it actually breaks). It is also the assumption of the design of PM tasks that when a problem is detected during inspection, scanning, etc. the maintenance system will respond with a corrective action. Activities you might find on a task list:

Life extension:

Clean

Empty

Tighten

Secure

Component replacement

Lubricate

Refill
Top-off

Perform short repair

Detection:

Inspect

Scan
Smell for...
Take readings
Measure
Take sample for analysis
Interview operator
Look at parts
Jog
Review History
Operate
Write-up deficiency

Common PM tasks:

Type of task	Example
1. Inspection	Look for leaks in hydraulic system
2. Predictive maintenance	Scan all electrical connections with infrared
3. Cleaning	Remove debris from machine
4. Tightening	Tighten anchor bolts
5. Operate	Advance heat control until heater activates
6. Adjustment	Adjust tension on drive belt
7. Take readings	Record readings of amperage
8. Lubrication	Add 2 drops of oil to stitcher
9. Scheduled replacement	Remove and replace pump every 5 years
10. Interview Operator	Ask operator how machine is operating
11. Analysis	Perform history analysis of a type of machine

These tasks are assembled into lists and sorted by frequency of execution. Each task is marked off when it is complete. There should always be room on the bottom or side of the task list to note comments. Items requiring action should be highlighted.

These tasks should be directed at how the asset will fail. The rule is that the tasks should repair the unit's **most dangerous, most expensive,** or **most likely** failure modes.

Caveat: There will still be failures and breakdowns even with the best PM systems. Your goal is to reduce the breakdowns to minuscule levels and convert the breakdowns that are left into learning experiences to improve your delivery of maintenance service.

In addition to Task List work the PM systems also include:

1. Maintaining a record keeping system to track PM, failures and equipment utilization. Creating a baseline for other analysis activity.

2. All types of predictive activities. These include inspection, taking measurements, inspecting parts for quality, and analysis of the oil, temperature, and vibration. Recording all data from predictive activity for trend analysis.

3. Short or minor repairs up to 30 minutes in length. Making such repairs is a great boost to productivity because there is no additional travel time.

4. Writing up any conditions that require attention (conditions that will lead or potentially lead to a failure). Write-ups of machine condition are included.

5. Scheduling and actually doing repairs written up by PM inspectors

6. Using the frequency and severity of failures to refine the PM task list

7. Continual training and upgrading of inspector's skills, improvements to PM technology

8. PM systems should contain ongoing analysis of their effectiveness. The avoided cost of the PM services versus the cost of the breakdown should be looked at periodically.

9. Optionally, a PM system can be an automated tickler file for time or event based activity such as changing the bags in a bag house (for environmental compliance), inspecting asbestos encapsulation, etc.

A special kind of failure: Hidden Failures

One of the toughest issues to deal with is the failure of a component that is hidden from the view of the maintenance department without special attention. These hidden failures require special tasks. The devices are usually protective in nature, and without special tasks many of the safety and protection systems cannot be verified as being in working order.

The simplest example is a temperature warning light in an automobile. Many vehicles only have a warning light and no gauge. How can you tell if your temperature warning light is burned out? This is true for a temperature/pressure relief valve on your hot water heater at home, the blowout port on the propane cylinder for your grill, and the switch in your heater that kills the fuel flow when the flame goes out. All these machines or components will happily operate without the protective device.

These devices have functions hidden from view in normal operation. The failure of a hidden function will not always be evident to operators or maintenance personnel unless they go looking for them specifically. In use, unless the car overheats, you might never know that there was a functional failure of such a hidden device.

Some of the biggest industrial accidents can be traced back to the failure of a single hidden protective device. It is essential that tasks be developed to verify that the protective device is on-line and ready to protect.

Six patterns of failure

Eventually, equipment, fleet vehicles, and buildings and their component systems require maintenance and all fail in fairly characteristic ways. These curves are called critical wear curves. Our maintenance approach and support systems (such as stores, computer support, engineering support, etc.) need to be sophisticated enough to detect which critical wear curve is most likely to be most typical of the asset's deterioration. Once the curve is selected we must locate where we are on the critical wear curve and act accordingly.

For example, if you would plot the failures (expressed in MTBF –Mean Time Between Failure expressed in hours of operation) in a thousand power supplies you would find a failure curve that looks something like curve 2 below. Most of the failures would occur in the first 48 hours. Subsequent failures are pretty evenly distributed over time. Note that if we actually look hundreds of years into the future we might find an ultimate life of these power supplies. Perhaps a component such as a capacitor will dry out in a century. The curve will then look more like curve 6 where the ultimate life is exceeded.

So far there is no effective inspection to catch a power supply on its way out (unless you are incredibly lucky). That does not mean that no PM is needed or effective. Life extension tasks might be keeping dust off the supply, keeping other dirt and debris out of the enclosure, replacing filters, maintaining adequate torque on connection screws etc.

Another type of asset such as the jaws of an aggregate crusher fail in a very different way. These jaws wear out and have to be changed or flipped over after many thousand tons of rock. The curve for this kind of equipment (designed to wear) is curve 3. In curve 3 the probability of failure gradually increases over time. It is very predictable and unlike a power supply, fairly well correlated to usage.

How to read the curves

Assets fail in different ways. These different ways can be expressed as curves. The trace of the curves represents the probability of failure over time. In all six curves, elapsed time or time of equipment utilization flows to the right. The probability of failure increases as the curve gets further from the X-axis. It is important to notice whether the curve representing failure probability is getting higher (away from X-axis), lower (toward the X-axis) or is stable (flat).

One complication is that every component system on each asset is on its own deterioration/ failure curve. The electronics, belts, motors, gears, sensors and everything else are all deteriorating in different patterns. The goal is to either decide which curve to use or look at the curves individually for major components.

All these curves have phases. The phase of interest to our organizations is called the wealth or use phase. In this phase the asset is used and wealth is derived from it (hence the name wealth phase). Generally the curve is flat and the failure rate is either low, predictable, or both.

The second phase of interest is how the asset starts up after installation, which is called the start-up or infant phase. This phase is usually the "fault" of the project managers but rapidly becomes the problem of the maintenance department.

The last phase is called the end of life or breakdown phase. What happens to assets in this phase is frequently of intense interest to maintenance departments. Many organizations trade or dispose of equipment when it reaches this phase. We see that high technology equipment usually becomes obsolete before it reaches this phase.

1- Random: The probability of failure in any period is the same (the probability of failure in month 109 is the same as the probability of failure in month 23). Failure can be caused by freak or random events. This curve is common for assets that don't wear out in the conventional sense or where you will keep the asset a short time (in relation to its life).

Example: A vehicle windshield will crack (fail) when a pebble hits it. The probability of a failure is unrelated to life span (unlike the gasket around the windshield which gradually wears out over time). The windshield does not wear out in the traditional sense. This random curve is common in electronics and in systems that become obsolete before they wear out (because the curve looks flat during the period of interest. Even with the windshield, a hundred years or a thousand years might see an upward tail to the curve.

2- Infant mortality: The probability of failure starts high then drops to an even or random level. This is a very common curve.

Example: Many electronics systems fail most frequently during the initial burn-in phase. After this initial period the probability of failure from period to period doesn't significantly change (becomes like curve #1). Most complex systems of any type have high initial failure rates due to defects in materials or workmanship. Manufacturers recognize this phenomenon and write warranties that cover most of these failures.

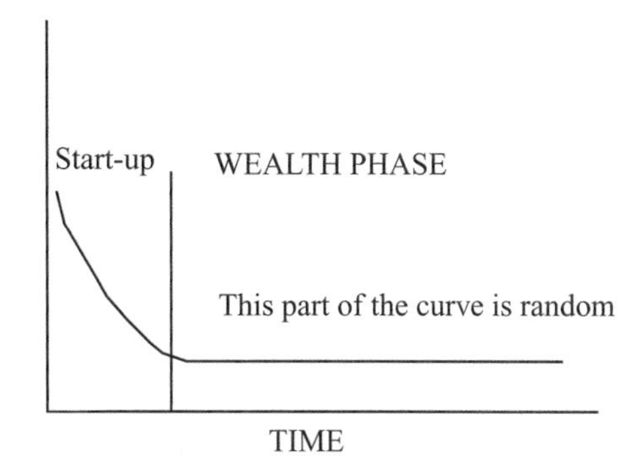

3- Increasing: The probability of failure slowly increases over time or utilization. This effect is common for items that are subject to direct wear. The curve shows no dramatic increase in failure rates. The engineer or skilled tradesperson determines where to make the change. Most change-outs occur after 67% or 75% of life.

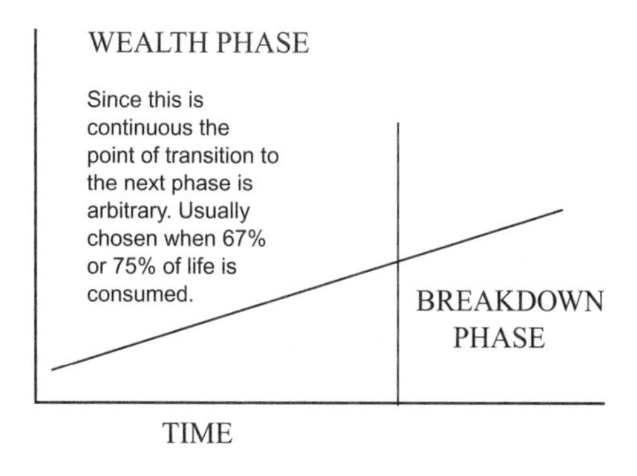

Example: Consider the jaws of an aggregate crusher. These are massive blocks of manganese steel that get worn away by the crushing action on the rock. They wear in a predictable way and the probability of failure increases gradually throughout their life. The curve will turn up if the item is allowed to wear too far. Systems that are changed at 67% or 75% will behave this way but might degrade to curve #5 if left too long. Most items subject to wear demonstrate a curve of this type.

4- Increasing then stable: The probability of failure increases rapidly, then levels off. This is not a common failure curve.

Example: An electric heating element in a hot water heater. The probability of failure increases as the unit is turned on and then stabilizes to a random level.

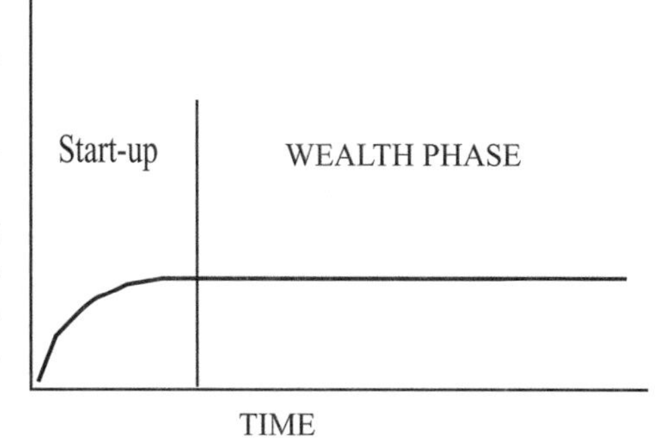

5- Ending mortality: The probability of failure is random until the end of the life cycle then it increases rapidly. This is a common curve configuration.

Example: This failure mode is characterized by mechanical systems that wear until they reach a certain point, after which they are at significant risk of failure. Failure modes related to corrosion usually proceed until the amount of metal left is marginal to support the structure. Failure rates dramatically increase when this level of deterioration is achieved. Truck tires on the trailer are changed by law at 2/32" tread depth. If the tires are changed as required the curve is quite flat (across between curve #1- random curve #3 gradually increasing) and no significant upturn is observed. If they are not changed, the failure rate turns up quite quickly and the whole system becomes unstable (and with tires, unsafe).

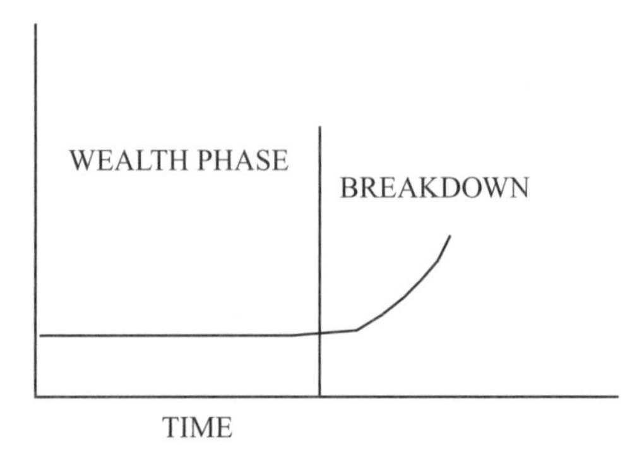

6- Bathtub: This curve is the combination of the infant mortality and the ending mortality curves. Probability starts high, then levels off, then starts to rise again. This curve is extremely common and is the only curve described in many maintenance texts.

Example: Trucks initially have high failure rates due to defects in labor and parts and intrinsic design flaws. Once these defects are eliminated, the vehicles fall into a flat section of the curve until one of the critical systems experiences critical wear. After critical wear occurs the whole reliability of the vehicle drops and the number of maintenance incidents increases until complete failure takes place. This is a common curve for complex systems that have start-up problems but are used until they are worn out.

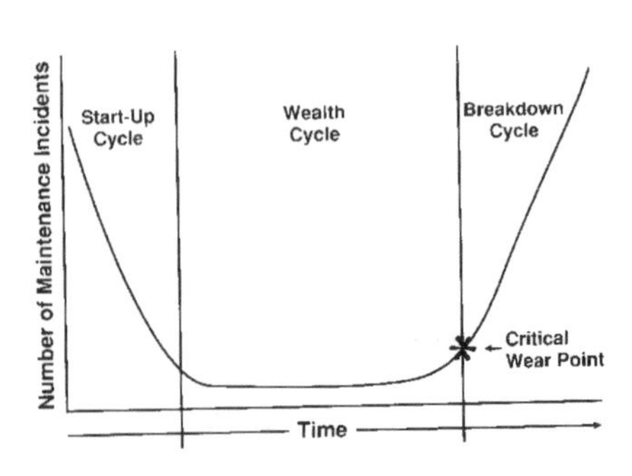

Proposition

The best strategy is closely related to the type of failure curve that is most responsible for the deterioration. The counter measures are designed by phase. Counter measures can be designed to either extend (as in the wealth phase) or minimize (as in the start-up or breakdown phases) the life. The counter measures will generally work for all curves that share a problematic phase (such as high failure rates during the start-up phase).

PM is part of this equation but not the whole solution. As you can see from the counter measures, good business practices also are important (such as leaving enough time to test run a new piece of equipment or product assembly line). PM cannot make an inadequate machine adequate but it can preserve and prevent a unit from deterioration.

1. Start-up phase. Infantile mortality. Represented most strongly by the infant mortality and bathtub curves. Failures of materials, workmanship, installation, and/or operator training on new equipment. Frequently the costs are partially covered by warranty. Unless significant experience exists with this make and model asset there is a lack of historical data. The failures are very hard to predict or plan for and it is very difficult to know which parts to stock.

The start up period could last from a day or less to several years for a complex system. A new punch press might take a few weeks to get through the cycle, and an automobile assembly line might take 12 months or more to completely shake down. Be vigilant in monitoring misapplication (the wrong equipment or machine for the job), inadequate engineering, inadequate testing, and manufacturer deficiencies.

Countermeasures: Excessive failures in Start-up cycle

Enough time to test run equipment properly
Enough time and resources to install properly
Operator training and participation in start-up
Pick the right equipment in the first place
Operator certification
Operator and maintenance department input into choice of machine
Maintenance and operator inputs to machine design to insure maintainability
Good vendor relations so that they will communicate problems
other users have
Good vendor relations so that you will be introduced to the engineers
behind the scenes
Maintenance person training on the equipment
Maintenance person training in start-up
Latent defect analysis (run the machine over-speed,
see what fails and re-engineer it)
RCM analysis to design PM tasks and re-engineering tasks
Rebuild or re-engineer to your own higher standard
Formal procedures for start-up (possibility of videotape?)

*2. **Wealth phase**.* All curves have a wealth phase (except where the asset is not strong enough for the job, then it goes directly from start-up problems to the breakdown cycle). The bathtub, infant mortality, random, and ending mortality each has a well defined middle phase. This cycle is where the organization makes money on the useful output of the machine, building or other asset. This can also be called the use cycle. The goal of PM is to keep the equipment in this cycle or detect when it might make the transition to the breakdown cycle. After detecting a problem with the machine or asset a quality oriented maintenance shop will do everything possible to repair the problem.

After proper start-up the failures in this cycle should be minimal. Operator mistakes, sabotage, and material defects tend to show up in this cycle if the PM system is effective. Also PM would generate evidence of the need for Planned Component Replacement (PCR). The wealth cycle can last from several years to 100 years or more on certain types of equipment. The wealth cycle on a high-speed press might be 5 years and the same cycle might span 50 years for a low speed punch press in light service.

Countermeasures:
Wealth cycle (designed to keep the asset in this cycle)

PM system

TLC- tighten, lubricate, clean

Operator certification

Periodic operator refresher courses

Close watching during labor strife

Audit maintenance procedures and checking assumptions on a periodic basis

Autonomous maintenance standards

Quality audits

Quality control charts initiate maintenance service when control limits cannot be held.

Membership in user or trade groups concerned with this asset.

*3. **Breakdown phase.*** This phase is best represented in the bathtub and ending mortality curves. The increasing curve also has a breakdown phase but it is harder to see where it starts. Organizations find themselves in this cycle when they do not follow good PM practices. The breakdown phase is characterized by wear-out failures, breakdowns, corrosion failures, fatigue, downtime, and general headaches.

This environment is very exciting because you never know what is going to break, blow out, smash up, or cause general mayhem. Some organizations manage life cycle three very well and make money by having extra machines, low quality requirements, and tolerance for headaches. Parts usage changes as you move more deeply into life cycle three. The parts tend to be bigger, more expensive, and harder to get.

The goal of most maintenance operations is to identify when an asset is slipping into the breakdown phase and fix the problem. Fixing the problem will result in the asset moving back into the wealth phase.

Countermeasures: Breakdown cycle

PM system

Maintenance improvement

Reliability engineering

Maintenance engineering

Feedback failure history to PM task lists

Great fire fighting capability

Superior major repair capabilities

Great relationships with contractors who have superior repair and rebuilding capabilities

Trade equipment before breakdown cycle (very popular in some industries)

Computerized Maintenance Management Systems (CMMS)

The CMMS (in its best form) is an integrated system that helps the maintenance leadership manage all aspects of life in the department. The area of concern in this chapter is the PM and PdM sub-sections or modules. If you review CMMS literature you will see that all systems offer some kind of PM module.

In *Maintenance Planning, Scheduling and Coordination* (by Nyman and Levitt published by Industrial Press) we discuss the CMMS as more properly a CMMIS (Computerized Maintenance Management Information System. Such a system provides information but does not directly help manage the department. In the PM subsystem we are clearly using the CMMS to help us directly manage PM. So the use of the acronym CMMS is clearly justified by the way we use the CMMS.

You could open any chapter in this book and read something about CMMS. In fact, everything in this book applies to the PM module of the CMMS. If you look at the diagram at the beginning of chapter 2, your PM system should allow you to run your department using that model. The system should support all the structures that are discussed here.

We have found that while all vendors have PM modules they are not all the same. Not all of are systems easy to use, or complete.

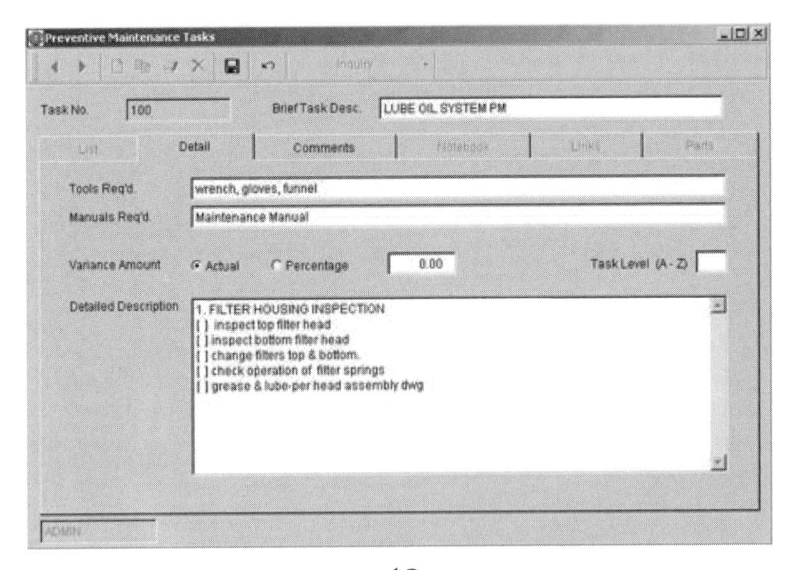

These screens are from the eMaint system, and have some interesting features.

One of the first steps is to have a place to enter the detailed tasks and planning data for the PM. In this example from eMaint three of the 6 elements of planning are listed (scope of work, tools, specialized information. Two additional items (trade required and standard hours) are on the next screen.

There are steps to setting up a PM program on a computer. We presume that the asset itself is already in the asset list. Once the PM task list is designed on paper, clocks are chosen and frequency is assigned it is time to build the CMMS files.

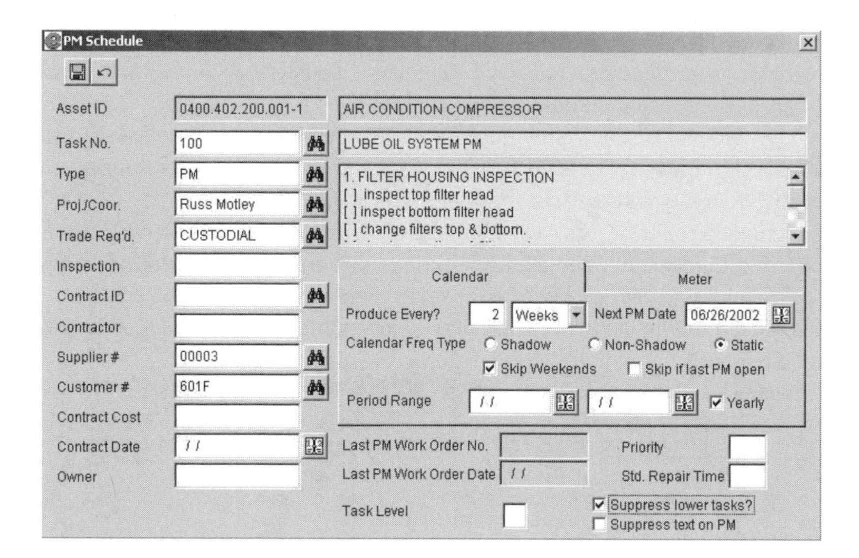

There are many good and even more adequate CMMS from a PM perspective. PM is important but it is a mistake to choose a system from that perspective only. In this book, several good systems are featured but they are used only as examples. There are other great systems that are not mentioned.

After the PM is created in the first screen, this second screen helps the PM manager schedule and define the task. It is clear that setting up PM is much more complicated than just tasks and frequencies. Some of the esoteric (but important) questions include:

1. Meter or calendar based?
2. Shadow, non-shadow, or fixed frequency? I had to ask what this meant. The issue is do you want PMs generated even if the prior one is not complete? The translation is, do you want the next PM to be generated 30 days after the prior one is generated or 30 days after the prior one is closed out or on the first of the month regardless?
3. Do you want PMs on weekends?
4. Do you want PM suppressed during a particular season?

A unified way to look at a potential system

All CMMS are designed with four major sections or functions. It helps to separate these functions and view them one at a time. The PM module integrates into each of these functions in a specific way.

Daily Transactions: Includes all data entry such as work order, receipts of parts, payroll information, fuel logs, and physical inventory information. For the PM, daily transactions are primarily completed tickets to close out outstanding PMs. Because the PM ticket (in some systems) is also used to capture data about short repairs and corrective maintenance work to be scheduled it should be designed to make it convenient to record these manual extras. In some instances the PM ticket will have readings and comments. One challenge is that mechanics sometimes make sketches or diagrams to explain things.

In this part of the system look for: completeness, speed of data entry, logical and consistent format. The ability to add in meter readings is important if you use that clock anywhere in your system.

There should be easy ways to add short repair and corrective data. Entering readings and mechanic comments should likewise be easy. The process of PM ticket closeout should be very quick. Most systems have only primitive capability to capture scanned diagrams or sketches and make them available to the next technician.

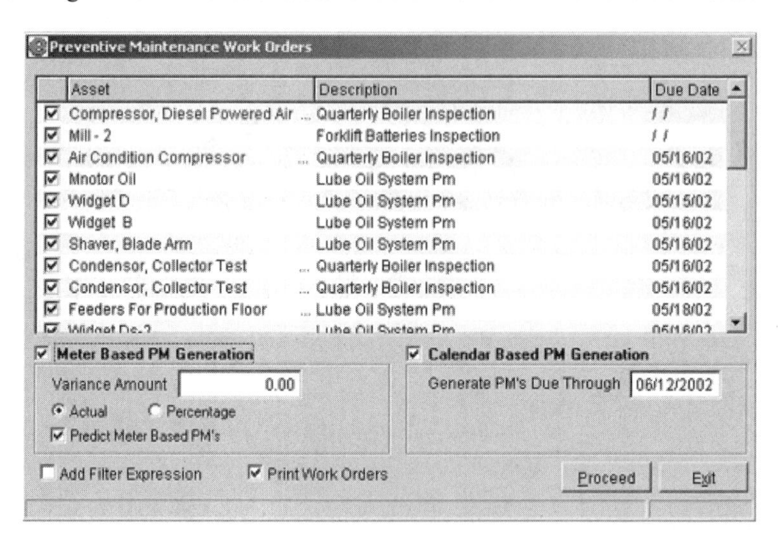

One important part of the daily transaction part of the system is work order creation. How difficult (or easy) is it to generate your PM work orders? The accompanying shot is of a PM work order creation screen. All PMs for the future period are displayed, and you can uncheck (not generate) any PM or group of PMs. One attribute that is interesting is that this system (eMaint) in resources section) will predict the meter reading for meter based PMs.

An organization will spend 10-20 times more money on this part of the system than in any other. In viewing this part of the system, look closely at portable solutions (PDAs, notebooks, laptops, and specialized processors that are common in the predictive maintenance world) both wireless and docked.

```
06/12/02                          PM Tasks                              Page # 1
--------------------------------------------------------------------------------
Asset                        Asset ID                                  Next PM Date
Produce PM Every Produce PM Every
Brief Task Desc.                     Last PM Work Order Date   WO No.
Work Description                                               Task No.
Task No.
--------------------------------------------------------------------------------
                             FEEDERS                W09/01/01      / /
    0              0
INSPECT AND LUBRICATE FEEDERS                                    49
Test                                                           FILINS0060
FILINS0060

                             010                    D 06/12/02   06/13/02
    0              0
LUBE OIL SYSTEM PM                                                 62
Task #: 100 - LUBE OIL SYSTEM PM                               100
100FILTER HOUSING INSPECTION
[ ] inspect top filter head
[ ] inspect bottom filter head
[ ] change filters top & bottom.
[ ] check operation of filter springs
[ ] grease & lube-per head assembly dwg

Tools / Parts Required : (Item \ Description \ Type \ Quantity)
MBLGRS-102    | LUBRICANT, FOOD GRADE      | Part |   1 Reserved
===================================================================
```

These advances can cut the time for data entry and improve accuracy. However the program works in the system of your choice, be sure the procedures are thought through fully and crewed. It is important that the daily flow of information be unimpeded and prompt.

Master files: The master files are the fixed information about the machines, buildings, vehicles, parts, mechanics, and organization. A full system might have 30 files of this type. New programming techniques and increased computational horsepower reduce the need for so many files but the effect is the same. Many different types of fixed information are needed to run a maintenance department. To a large degree the master files determine what analysis can and can't be done by the computer. When they are complete and accurate, these files are useful in themselves.

In the PM module the master files include the tasks, frequencies, and hierarchy. The "PM" might be directly in the equipment file (not very flexible) or be a separate set of files by itself. The master file for the equipment also shows to which PM class (like equipment in like service) the asset belongs .

Look for: Completeness in the original design. Completeness is important because it is very difficult to add any fields to a master file after it's in use. Not having space in a master file for information that you want to store (perhaps you discover the gap after the system is in use) is a major difficulty.

PM has specific requirements for master files that depend on your need and use of the system. For example, the type of clock that initiates the PM has to be managed carefully. Some machines might use a calendar, some a kilowatt-hour meter, others an hour meter, and still others product output (like tons of product). The master file structure has to track and not mix or confuse the units from different clocks. If you plan to just use the calendar based PM then a shortcoming in this area is not fatal.

In the accompanying screen notice there are 7 or 8 tabs of information for the asset. For this example, the asset files can be 4 levels deep (system, sub-system, component, and sub-component).

There are several structures of Master Files that make the job of managing PM useful. One of the best structures is a master file for PM that stands alone where you can make one change and data for all the equipment in the class will change. Another is the ability to clone a task list for a similar asset.

Processing: The daily transactions are processed to the operational files. Processing updates the PM schedule, summarizes detailed repair data for reports, and keeps all accounts current.

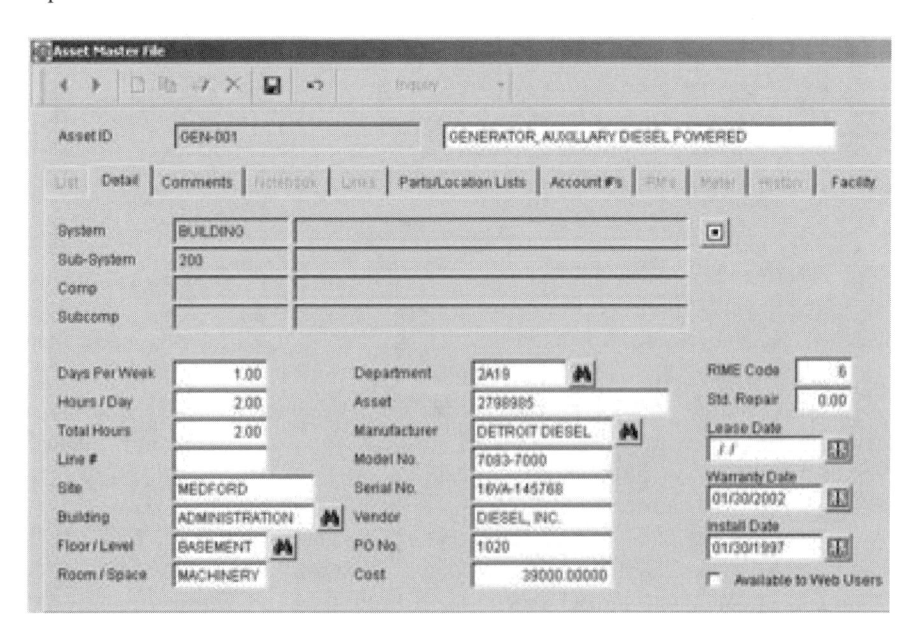

Look for: Does it work? Take the time to process some data through the full cycle and see if all the accounts, schedules, and master files are updated correctly. Accuracy and completeness are the difficult areas here. Most of your bugs will occur during unusual processing conditions.

The PM module does not require severe use from the processing program. When groups of PM tickets are returned they are entered into the system. The system resets the clock for each PM completed. If needed (and set up) the program will also reset the clock for PM lower on the hierarchy. Completion of an annual PM would reset the monthly PM clock, avoiding excessive PM and duplicate services.

Demands: The demands on a maintenance system include printed reports and screens. Printed reports are needed where there is a large amount of data or where analysis is required. Inquiries should not have to require going to the print. Imagine how you expect to use the system and then see how the system will behave.

Look for: Many different ways to look at the data, complete basic sets of reports and screens, and future ability to alter or add reports/inquiries to suit your changing needs and growing expertise. Most current systems have sophisticated report writers either built in or available as add-ons. PM has some particular kinds of reports that really make PM easy to manage. These reports help at specific times such as PM design, PM scheduling (called loading), and on-going PM management.

Reports in the PM design phase include unit histories sorted by component, class histories (same as unit history but for a class of equipment), MTBF (Mean Time Between Failures) by component, and reports or listings of notes by mechanics who worked on the equipment.

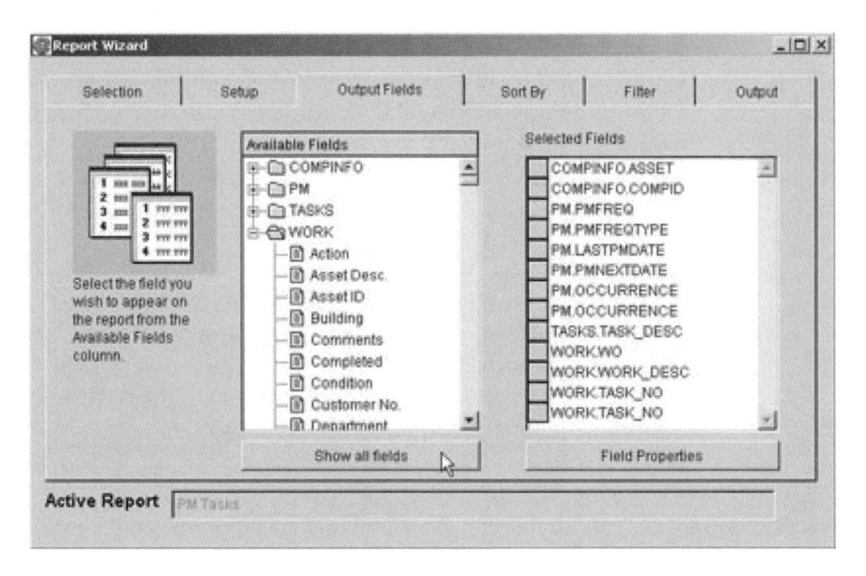

The loading phase may generate reports that predict the PM requirements for an entire year in advance, reports that summarize labor requirements for the upcoming week-month-year, and reports that list material requirements for various periods.

To manage PM the supervisor needs to know PM compliance, PM statistics (hours, tasks complete, etc), breakdown reports, and breakdown trends (showing how effective PM has been).

Thoughts on installing a PM program

If you are installing a PM module for the first time by all means read the checklist in Chapter 20 for all the steps necessary. When looking at the system itself, here are some questions to discuss within your PM taskforce:

1. What do we need for our particular operating pattern? In other words is a basic PM with a calendar adequate or do we need something more sophisticated?

2. Is the system we own or are looking at adequate for these requirements? Does it have enough flexibility to meet our needs?

3. Does the system we are considering require major changes to our existing procedures, or to our command and control system?

Shop floor automation

The increasing levels of sophistication of shop floor systems has made it easier and easier to get operational data into the CMMS. New condition-based systems also are starting to communicate with CMMS. (From the Instrument Society of America (ISA) June 2002 magazine)

Level 0	Conventional and or intelligent instruments and actuators
Level 1	Local control units (LCU) such as PLCs and intelligent I/O. Communicates with level 0 through 4-20ma, dry contacts or field bus networks.
Level 2	Area Control (AC) usually PC workstation networks with attention to operator interface
Level 3	Communication and Information system (CIS) sis where functions of the company such as accounting, CMMS, MRPII interface. There is just beginning to be the ability to start in the CMMS and drill down through the AC to the LCU to find out hour meter readings and temperatures.

Increasingly modern SCADA systems, building management, and process control systems (at the AC level) are providing real time data to the CIS level.

Short Repairs and High Productivity

From the glossary, Short Repairs: *Repairs that a PM or route person can do in less than 30 minutes with the tools and materials that they carry. These are complete repairs and are distinct from temporary repairs.*

Short repairs are done at the time that the PM inspector is looking at the asset. These are repairs that can be done without going to the stockroom, provided the mechanic is already carrying the tools needed to do them.

There is an extremely persuasive argument in favor of incorporating short repairs into your policies and procedures. Maintenance worker productivity is traditionally pretty low. For every hour you pay for, the average maintenance worker does less than 20 minutes of direct maintenance work (based on extensive work sampling in numerous maintenance situations). The question for the manager is: what is the person doing when they are not doing maintenance? The second and more important question is: what can a manager do to the maintenance process to improve this situation?

Notice the use of "the maintenance process." We did not say what action could the manager take against the supervisor or worker. In all the studies, the actions of the maintenance worker were quantitatively less important than the process in which they operate. In fact there is significant anecdotal evidence that the process produces the negative attitudes occasionally seen in maintenance workers.

By process we mean the steps needed to get work done. So the process can be defined as what the maintenance worker goes through to get materials, get job assignments, take over control of a machine, etc. It is these processes that doom maintenance to low productivity, and it is the processes that must be changed.

The results of work sampling diagrammed below report on the result of the process. This report card is an important measure of how the process is doing. For more details on work sampling techniques consult *The Handbook of Maintenance Management* by Joel Levitt, published by Industrial Press.

Given the state of maintenance productivity, why are short repairs so important? The answer is that short repairs can be a simple change to the maintenance process that results in a major change to the productivity.

Out of an average 480-minute day the maintenance worker might be expected to spend:

Breaks and excess personal time	113 minutes
Travel (transporting tools and materials)	77 minutes
Idle (no job assignment)	44 minutes
Waiting (for unit, for permit, etc)	22 minutes
Cleaning up tools (between jobs, afternoon)	25 minutes
Getting assignments	21 minutes
Late to work/left early	21 minutes
Direct maintenance work	**157 minutes**
	480

Create the scene in your mind. A maintenance worker is at a machine to perform a PM service with the task list in his or her hand. What did he or she have to go through to get to that point?

To start with they got the job assignment to PM that particular machine. They went to the supply room and picked up any materials necessary for the job. They had to travel to the location of the machine. On arrival at the machine they had to wait to get control (or not) and lock out the machine (if that was necessary). Then they are ready to start work. If you view the non-productive time as overhead then the PM ticket for that machine has already funded the overhead. **Any additional work done at that time is pure productivity**. Saying it another way, adding a mere 16 minutes of short repairs into an average day adds about 10% productivity to that day!

There are three rules for short repairs

1. You have to set a maximum time depending on the size of your facility and your type of equipment. Usually the limit is 15 minutes but limits as long as an hour are common.

2. The repair must be capable of being done safely with the tools that the PM person has on hand.

3. The last rule is that the PM mechanic must be already carrying any necessary materials or parts

Design a cart

Investment in thinking through the contents of the PM-Short Repair Cart is time well spent. Each kind of asset group might need additional small parts and an occasional specialized tool. The size, shape, or the cart can vary from a 5-gallon bucket to a large truck/trailer combination with shelves and a mini-shop.

A special argument is made in favor of facilities remote from your main operation. A northern Canadian natural gas firm designed their PM-Short repair truck with enough tools and spares to almost rebuild their remote pumping stations. Of course these computer-controlled stations were not manned and could be 300 miles from the nearest settlement! If a PM inspection showed that there was an impending failure you can bet the short repair could last 2 or more days.

Short repairs in housing, office buildings, and other facilities are essential for even minimal levels of productivity. Most jobs in these kinds of maintenance situations are small and can be handled by effective short repairs. To maximize short repair possibilities in an Apartment Building the PM maintenance person might be equipped with:

1. Hand tools including: (screw driver set, pliers set, claw hammer, cutters, hexagon wrenches, vice grips, key hole saw, hack saw, tape measure, utility knife, pipe wrenches, set of files, rasps, good flashlight, batteries etc.), stepladder to reach ceiling

2. Electric tools such as: electric drill and bits, drop light, Skil saw (battery powered is great, otherwise carry 100' extension cord, 3 prong adapter)

3. Cleaning tools (Straw broom, whisk broom, dust pan, trash bags, mop, wringer, bucket, pick-up stick with nail end, rags, shovel, sponges, 5 gallon bucket, spray bottles, razor blade scraper, steel wool)

4. Cleaning supplies (furniture polish, all-purpose cleaner with TSP, spray deodorizer, spray tile cleaner, wax, wax applicator, wax stripper, toilet bowl cleaner, oven cleaner, metal polish, non-abrasive cleanser), rags, paper towels

5. Silicone spray lube, WD40, spray paints, spray zinc, standard off white latex paints (or standard colors) with brushes and rollers, joint compound, spackle knife, spackle tape, contact cement, latex and silicon caulk and gun,

6. Variety of packs of fasteners, variety of nails, small hardware items, duct tape

7. Electrical: light bulbs (of suitable wattage), florescent replacement tubes, switches, outlets, switch, outlet and blank covers, electrical tape, fuses, fittings, outlet tester, neon tester, door hardware, lock sets, door bells, transformers, bell wire, smoke detectors, batteries, tags for writing dates of installation and testing

8. Window hardware, floor and ceiling tiles, threshold and entrance strips

9. Bug bombs, insecticide spray, can hornet/wasp killer, roach/ant traps

10. Faucet washers and seats (seat tool), kitchen and bathroom faucets with flexible lines

11. Toilet parts, closet seals, toilet seat parts, closet snake

Adding to the PM cart

1. Each cart or each area should have a Cart Inventory list. The cart should always carry these items. It is important that the last daily task is to replenish and clean the cart.

2. Study the maintenance log and the corrective work orders. Add items based on jobs requested.

3. Periodically meet with the PM crews and discuss jobs completed and jobs that could not be completed. Adjust the cart list based on these discussions.

4. Allow the individual PM personnel to add things to the cart on their own. Again, at the periodic meeting discuss the individual additions to see if they warrant adding to the cart inventory list.

The key to these carts is discipline. The tools and unused materials are put away into the same places, pockets, drawers, and cabinets each time they are used. Care is taken to clean, lubricate, charge batteries, and generally care for the tools every PM day. There is nothing more frustrating then being in the middle of a short repair and having a dead battery on a needed screw gun, meter, etc. Replenish the parts used. For the records, make a log entry (or if possible, add a written line to the bottom of the PM ticket). Note on it what was used.

Tremendous thought goes into how to outfit a service person's truck. Next time you have an opportunity ask the telephone installer or gas repair person how their truck is set-up and why. The more often they have the needed part with them the more money you save. When they can use items from stock they take the best price rather than the local neighborhood hardware store price (or waste time with P.O.'s and supply companies).

Route Maintenance

In a building, large facility, or sports venue there is a variation on short repair that could even apply to a factory or even a fleet. It's called route maintenance

Develop a route so that you visit every part of every property and every unit every month (in a more active building, a weekly route might be better). Where appropriate, you could combine this with your PM walk through. Notify the tenant/user in advance to make a list of minor items (or put a box near the door of the building for quick write-ups). Try to hit each area on the same day and time of day. Route maintenance is a great way to deliver excellent customer service at low cost. As the route maintainer walks through they note and repair other small items that are seen. .

The route maintenance person needs to be multi-skilled so that they have a good shot at correcting any problem that may be encountered. Anything that they can't handle should be written up on a standard write-up form. That form is put back into the CMMS to become plannable workload. Of course, any emergencies need to be processed and scheduled immediately.

Route maintenance is most effective if your mechanic can fix everything they encounter without a trip to the storeroom or hardware vendor. As the route is traversed and a history is developed, (and log sheets filled out) the workers will get better and better at equipping themselves. Use the log sheets and look at the most common minor problems encountered in the last six months or year. From this list and using your own experience, create a list of tools and materials.

Reliability Enhancement Programs

FMECA

One of the earliest techniques for improving reliability is FMECA. FMECA stands for Failure Mode, Effect, and Criticality Analysis. The FMECA.com site says "The FMEA discipline was developed in the United States Military. Military Procedure MIL-P-1629, titled *Procedures for Performing a Failure Mode, Effects and Criticality Analysis,* is dated November 9, 1949. It was used as a reliability evaluation technique to determine the effect of system and equipment failures. Failures were classified according to their impact on mission success and personnel/equipment safety."

As with any good idea, software is available from many vendors. The Haviland Consulting Group and its software partner Kinetic, LLC have combined to offer a program to facilitate the FMEA process, known as FMEA Facilitator. This package was priced from $1100 for a single user to over $100,000 for entire corporations (see www.FMECA.com in resource section).

FMECA of a complete machine is a time-consuming exercise as is (RCM), especially since the various causes of breakdowns on the different parts of a machine may be interdependent. Thus, use of FMECA should be restricted to assemblies and sub-assemblies, and only in exceptional cases should it be extended to components. This aspect is related to RCM in that it is a series of techniques designed along the same lines but includes some systematizing in areas where RCM doesn't. Some of this material is adapted from *http://www.maintenance-tv.com/servlets/KSys/129/View.htm* (see resource section)

In the first part of an FMECA study, you

Make a list of the sub-assemblies / components
List the failure modes
Study the possible causes of failure
Find out how possible failures can be detected

In the second task, we will rank the different failures based on these data and decide what type of maintenance to perform. The choice of which type of maintenance will be performed is based on a ranking. Three factors are taken into account:

Frequency: how often does a failure occur?
Gravity: if a failure occurs, how serious are the consequences?
Detectability: ease or timeliness of detect ability

The criticality is the product of these three factors, and provides a ranking of the failures. Finally, we decide upon the right type of maintenance based on these figures (Corrective Maintenance, condition based, time based, and Design-out Maintenance (re-engineering).

In the third part, we will define the right actions to take based on these outcomes

1. Recommended action per failure
2. Define frequency of action. This frequency is based on the PM clock and is directly related to the rate of breakdown
3. Choose to act while the machine is running or shut down (This aspect can be considered one version of interruptive versus non-interruptive maintenance. The choice has to be made whether to break in to the production cycle or not.
4. List responsible persons (is the task to be done by maintenance, operations, or others)

Special use of terms for FMECA from FMECA.COM

Cause: A Cause is the means by which a particular element of the design or process results in a Failure Mode.

Criticality: The Criticality rating is the mathematical product of the Severity and Occurrence ratings. Criticality = (S) * (O). This number is used to place priority on items that require additional quality planning.

Current Controls: Current Controls (design and process) are the mechanisms that prevent the Cause of the Failure Mode from occurring, or that detect the failure before it reaches the Customer.

Customer: Customers are internal and external departments, people, and processes that will be adversely affected by product failure.

Detection: Detection is an assessment of the likelihood that the Current Controls (design and process) will detect the Cause of the Failure Mode or the Failure Mode itself, thus preventing it from reaching the Customer.

Effect: An Effect is an adverse consequence that the Customer might experience. The Customer could be the next operation, subsequent operations, or the end user.

Failure Mode: Failure Modes are sometimes described as categories of failure. A potential Failure Mode describes the way in which a product or process could fail to perform its desired function (design intent or performance requirements) as described by the needs, wants, and expectations of the internal and external Customers.

FMEA Element: These elements are identified or analyzed in the FMEA process. Common examples are Functions, Failure Modes, Causes, Effects, Controls, and Actions. FMEA elements appear as column headings in the output form.

Function: A Function could be any intended purpose of a product or process. FMEA functions are best described in verb-noun format with engineering specifications.

Occurrence: Occurrence is an assessment of the likelihood that a particular Cause will happen and result in the Failure Mode during the intended life and use of the product.

Risk Priority Number: The Risk Priority Number is a mathematical product of the numerical Severity, Occurrence, and Detection ratings. RPN = (S) * (O) * (D). This number is used to place priority on items than require additional quality planning.

Severity: Severity is an assessment of how serious is the Effect of the potential Failure Mode on the Customer.

Significant Characteristics: Significant Characteristics are Special Characteristics defined by the customer as characteristics that significantly affect customer satisfaction and require quality planning to ensure acceptable levels of capability.

Special Process Characteristics: Special Process Characteristics are process characteristics for which variation must be controlled at some target value to ensure that variation in a Special Product Characteristic is maintained at its target value during manufacturing and assembly.

Special Product Characteristics: Special Product Characteristics are product characteristics for which reasonably anticipated variation could significantly affect a product's safety or compliance with governmental standards or regulations, or is likely to significantly affect customer satisfaction with a product.

The RCM (Reliability Centered Maintenance) approach to PPM

RCM has made great contributions to the whole maintenance field. Originally RCM was designed as a process to insure reliability in aircraft. The ideas and tenets have filtered onto the shop floor, into the fleet garage, and into our facilities. Formal RCM investigations require a structured approach to look at the functions, functional failures, failure modes, and consequences on a component-by-component basis.

These studies can cost $1M or more (when you include the required fixes). Formal RCM is carried out only in the largest organizations with significant maintenance costs or severe consequences for failure (or both). Smaller organizations with particularly severe consequences can also benefit. The costs of these studies are related to the complexity of the equipment and the amount and depth of the re-engineering necessary.

There are many elements of RCM that can be adapted by even the smallest companies, and all maintenance entities can take advantage of the ideas in RCM. These ideas were revolutionary in the aircraft business and now 40 years later are evolutionary in the general maintenance world.

1. The consequence of the failure is more important than the failure itself. In RCM a significant amount of effort is invested in cataloging the consequence of various failure modes. The reason is because the way that a potential failure is treated is based on the levels of losses (both financially and human). The same valve will be treated differently under RCM if in one example it is a critical component in an intensive care unit (hospital) and in another, a component of a swimming pool system.

2. If the consequence of the failure is death or significant environmental damage then the risk has to be either designed out of the system or PM tasks designed so that the probability of an unanticipated failure is close to zero. What is interesting about the RCM approach is that before RCM was invented the engineering profession would eventually go through a vaguely similar process but at a much slower rate. Whenever there has been a dangerous failure mode both society and conscientious engineering forced redesign until failure rates were reduced to "acceptable" levels.

3. For failures that cause downtime and repair costs, design a PM task that costs less to perform annually than the cost of the failure, plus the cost of the downtime, times the probability of the failure recurring that year. In other words the cost of the task has to be less than the cost of the failure.

4. Some of the other benefits include the process of thinking about your equipment in a structured way. The advantage of this approach can be seen in troubleshooting (you can use the RCM diagrams), replacement decisions (look for unintended consequences), and in designing task lists.

One of the most important aspects of RCM is its holistic approach to equipment. The equipment is viewed from many different vantage points including operations, maintenance, storage, energy sources, and others.

Think for a minute about the issues surrounding RCM and aircraft. Looking at the engine system imagine listing all of the possible failure modes associated with the loss of function called inadequate thrust. Not being a jet engine expert I can only imagine the failure modes. The most interesting for this discussion is that RCM would consider problems with fuel quality or contamination, inadequate quantity of fuel, improper resetting of the speed control by the pilot and other modes not related to maintenance efforts.

RCM portends a breakthrough in thinking and can be applied to all maintenance situations. In your own organization, how often is loss of function associated with something the operator or the public did?

A second major contribution is the distinguishing of levels of breakdown. In RCM breakdowns (they are referred to as loss of function) are divided into three levels or grades by the consequence of the breakdown. These levels are:

1. Breakdowns where the consequences are loss of life or environmental contamination such as a stuck boiler safety valve or the rupture of a tank of volatile chemicals.

2. Failures where the consequences are operational downtime such as loss of cooling water to a data center or the breakage of the drive chain in an auto assembly line.

3. Failures where the consequences are repair costs, such as the breakdown of one of several milling machines in a machine shop.

For failures with severe safety and environmental impacts (Level or grade 1)

The only acceptable alternative for a category one failure mode is redesign to eliminate that mode or assignment of a PM or PdM task that reduces the probability of that failure mode to almost zero.

One of the most dangerous jobs in the beginning of the twentieth century was boiler operator. In any major city, a boiler explosion and associated deaths and destruction of property were a weekly occurrence. Over a two or three decade period the engineering profession responded with improved specifications on manufacture, licensing operators, blow-out ports, pressure-temperature relief valves, low/high water cut-offs, flame detectors, and more rigorous operating practices. These changes occurred well before RCM but they reflect an RCM type of approach. In the present era, boiler explosions are quite rare and when they result in loss of life are considered front-page news.

In aircraft this attitude shows up in equipment redundancy, licensing of mechanics (specific to an aircraft), certified rebuilding programs, tight procedures, formal sign-off and turnover procedures, and massive data collection by the FAA to detect trends.

For category two and three

In categories two and three, the fix has to be justifiable based on the costs of the fix compared to the cost of the loss of function. If RCM analysis shows that an item needs a PM task to detect the impending failure then economics will be considered. Each task (line item) should be considered carefully before adoption because inclusion will create a cost for the long term. The quick way to evaluate task economics is to relate the task cost per year to the avoided cost of the breakdown. In category two or three, economic analysis is the way to determine if a task should be included. In category one, the task must be included or the asset re-engineered to remove the threat of failure.

Breakdown costs formula:

(Probability of failure in 1 year) * (cost of downtime + cost of repair) <

PM Costs formula:

(Cost per task * # services per year) + (New Probability of failure in 1 year * (cost of downtime + cost of repair))* 0.7

So a $50 (1 hour plus $10 of materials) weekly task for a failure mode with a 2 or 3 category consequence (no safety and/or no environmental consequences) has a hard cost of $2600 per year. This task can only be justified if the total fair cost of the failure is over about $3715 a year. By fair cost we mean a true cost without embellishment. It is somewhat more complicated than that because you have to add back in the new probability of failure, times the failure cost (as we did in Chapter 3 examples).

RCM Chart Asset Name:or ID# Page___of____

Date: Asset Decription:

Function	Loss of Function	Mode of Failure	E	S	O	N	Description of task or reengineer

RCM focuses on hidden functions

The hidden failures are a big part of the functional analysis in RCM. Analysis is always looking for protection systems and verification that they are in working order. Pressure relief valves, warning systems, shut down circuits, are all included in the RCM review. In such equipment as boilers, the protective device is removed and sent to a rebuild/certification shop on a periodic basis.

Important questions

According to the Maintenance-TV.com site (see resources) RCM is designed to ask and then answer these seven questions. The questions are simple but the inquiry to get to the answers may be profound:

1. What is the function of the equipment (or component)?
   How are its performance requirements measured?
2. How can the equipment fail to fulfill these functions?
3. What can cause each failure?
4. What happens when each failure occurs?
5. How much does each failure matter? What are its consequences?
6. What can be done to predict or prevent each failure?
7. What should be done if a suitable proactive task cannot be found?

RCM work sheet follows on the next page:

There were many additions to, quick versions of, and simplifications in early versions of RCM. Some of these techniques raise the ire of the 'pure' RCM practitioners. (One just has to read the trade press to see an emotional exchange).

PMO (PM Optimization)

What would happen if you took the good structures of RCM but skipped the function part and looked only at failures that have happened or are likely to happen? Well if you skipped these steps and added in some common sense you would end up with PMO. PMO is specifically designed for mature industries where the opportunity for equipment redesign is limited.

RCM came out of an environment where, if the system was a problem it could be redesigned. In most factories, buildings, and certainly fleets of vehicles, the equipment is just a given of the equation. Some redesign can be done, depending on the capabilities of the organization, but it is usually very limited in scope. Typically factories have more capability for reengineering than fleets or building maintenance departments.

I would like to thank Steve Turner, a professional engineer from Australia and RCM expert, for introducing me to PMO. He developed PMO out of a frustration with the application of RCM in mature industries. Much of the material is adapted from his writings. He can be contacted through pmoptimisation.com.au (details in the resource section).

Nine steps to PM Optimization

1. Task Compilation

Create a catalog of all tasks already performed by anyone who has contact with the equipment. Normally these people would comprise all current PM tasks (of course) but also tasks done by machine operators, quality personnel, cleaners, calibration departments, safety inspectors, and others. Every task should be listed with frequency. It is recognized that some of the PM effort is not documented and is carried out on an ad hoc basis by tradespeople, operators, and contractors. Thus, a compilation of the written task lists from the CMMS (if one is in place) can be a starting point but is not sufficient. Direct interviews are needed with all parties that come into contact with the equipment. Generally, the data is collected into a spreadsheet, which facilitates further steps with the ability to sort the data into different columns.

Task	Trade
Task 1	Operator
Task 4	Maintenance
Task 2	Operator
Task 7	Maintenance

2. Failure mode analysis

In RCM a great deal of thought and time goes into looking at the functions and the function failure engineering to determine all possible failure modes. In PMO failure mode analysis the team works from the accumulated tasks back to the failure mode. In other words, failure modes without tasks are not considered initially (they are considered later). This one difference cuts the time of the project dramatically over an RCM project of the same size. The task compilation is the basis for this part of the project. The question to be answered is, what failure mode is being addressed by each task? A cross-functional team is best for this kind of analysis.

Task	Trade	Cause
Task 1	Operator	Failure 1
Task 4	Maintenance	Failure 3
Task 2	Operator	Failure 2
Task 7	Maintenance	Failure 1

3. Rationalization and FMA (failure mode analysis) review

Rationalization is simple, put like causes together so that tasks addressed at the same cause are next to each other. Officially, if all the tasks from all the sources were loaded into a spreadsheet, then sort the spreadsheet by failure mode.

Task	Trade	Cause
Task 1	Operator	Failure1
Task 7	Maintenance	Failure 1
Task 2	Operator	Failure 2
Task 4	Maintenance	Failure 3
		Failure 4

At this point the team can readily see if there are failure modes covered by duplicate tasks or covered by clearly inadequate tasks.

Equipment history is consulted to make sure that all failure modes are listed. The team reviews the engineering for the asset as well as the asset itself, and determines whether there are significant failures that are not covered by any task. Hidden failures are frequently in this category. Failure four in the chart is a failure without a task associated with it.

4. Optional Functional analysis

In some analyses an RCM type functional analysis and evidence of loss of function are indicated. This approach can be justified on highly complex equipment where the consequences of failure are severe. In these few events, a sound understanding of function is essential to determining that all maintenance and operational issues are covered.

5. Consequences Evaluation

One of the breakthroughs of RCM is its focus on consequences rather than on failures themselves.

Task	Trade	Cause	Effect
Task 1	Operator	Failure 1	Operational
Task 7	Maintenance	Failure 1	
Task 2	Operator	Failure 2	Operation
Task 4	Maintenance	Failure 3	Hidden
		Failure 4	Operational

The failures are looked at for consequences. The consequences divide themselves into two logical categories of Hidden and Evident. A hidden failure is the burning out of a warning light on an instrument panel (the failure is not evident since the light is normally out). A further analysis of the evident failure modes looks at level of hazard and operational consequence. These categories are analogous to the ESON factors in RCM.

6. Maintenance Policy Determination

This step is the core of PMO. Based on the consequences, certain decisions are made for each task. There is a series of questions the PMO team asks about each task.

The first determination concerns microeconomics. Is this individual task (labor and materials times frequency per year) worth the cost, given the cost of the failure times the probability of the failure in that year?

Is there a better way to get to this failure mode? In some circumstances, introduction of quick condition-based monitoring would save overall time and money. The corollary is, would this task respond to simplification of the technology?

What tasks serve no purpose and can be eliminated? Along with those questions, which tasks can be set up at lesser or greater frequencies to increase effectiveness.

There is always an issue of what data to collect and to what end. The analysis at the task level answers this important question. What data should be collected to be able to predict the life of this component more accurately?

7. Grouping and review

This step is very practical in that it looks at the tasks that are left after duplicates and uneconomical work were eliminated and divvies them up based on the facts of your operation. Questions like: does operations get all the daily tasks; and should the night shift be given accountability for this specific asset; are answered.

8. Approval and Implementation

All parties have to be informed about what changed and why. All stakeholders are involved in this step. It is essential that the change be communicated to both maintenance and operations personnel and staff. Chapter 20 has details about this important step. The more complex the operation is, the more important this step becomes (and the more difficult).

9. Living Program

Turning a PM program into a living program requires time and patience. Less wasted PM will mean immediate freeing of resources. As these resources are reinvested to clean up the backlog, and the effective PM strategy takes hold, the number of breakdowns decreases. As the number of breakdowns decreases, more resources are freed up and can be used to accelerate the whole program.

Other steps can now be taken to improve the whole maintenance effort. In this context the changes now contemplated will make a difference.

PM Details for Effectiveness

Four Types of Task Lists

Unit Based: This is the standard type of task list where you go down a list and complete it on one asset or unit before going on to the next unit. The mechanic would also correct the minor items with the tools and materials they carry (called short repairs- Chapter 7).

Another variation of unit PM is Gang PM where several people (a gang) converge on the same unit at the same time. This method is widely used in utilities, refineries, and other industries with large complex equipment and with histories of single craft skills. Gang PM is also common in industries with scheduled shut downs (power utilities, automobile assembly, etc.)

A third variation of the unit based task list is the TPM approach. In a TPM run factory the operator is responsible for the unit PM. The TPM generated PM might be daily, weekly, or occasionally monthly. A mechanic might also be in the loop and be responsible for a more in- depth annual PM.

> Advantages: the mechanic gets to see the big picture, parts can be put in kits and made available from the storeroom as a unit, a person learns the machine well, the mechanic has ownership, there is a travel time advantage (only requires one trip), The worker gets into the mindset for the machine, and it is easier to supervise the worker than with other methods. The mechanic can discuss the machine with the operator as an equal partner.
>
> Disadvantages: high training requirements, higher level mechanics needed even for the mundane part of the PM, short repairs can put you behind schedule, and if PM is not done, no one else looks at the machine.

Different types of PM task lists may look very different visually. When you are looking at a new task list, imagine what failures each task is associated with. Also consider the cost of each task and the cost of the failure it avoids.

Many of us start the day with a good cup of coffee from one of the increasingly common coffee shops. These shops feature a complex espresso machine. Almost all the tasks on the task list below are related to quality assurance. If you get a couple of bad cups of coffee you will cross that shop off your list. Note that even in a small-scale retail operation, an economic analysis of this task list is still possible.

Frequency	PM routine for Espresso Machine adapted from programs of Espressoparts.com http://www.espressoparts.com/allabout/maint.html	Description
1. Q	Service filtration system (or outsource this item).	A water filtration system should be in placeRegenerate softener for most espresso machines.
2. W	Clean the group head.	Important- this is where the coffee comes in contact with the machine.
3. D	Back flush with water	Back flush the machine for about 15 seconds. The blind filter will cause the water to pressurize, and help to clean out any accumulations of coffee grounds and oils that may have formed.
4. W	Back flush with Purocaf or other NSF approved detergentImportant note: Do not back flush piston machines!Instead, just replace the screens and . gaskets on a regular basis	Afterwards it is important to remove the porta-filter and run the group again to rinse out all remaining detergent andback flush several more times with water only. In addition to rinsing, one or two shots of espresso should be extracted through each group to “re-season” the machine.
5. W	Soak porta-filters and screens in detergent (after back flushing)	Follow dilution instructions with very hot water. Be sure however to rinse portafilters well before re-using.
6. D	Clean group gaskets	Cleaning is best accomplished using a specially designed group cleaning brush and hot water to vigorously scrub around the sealing surface.
7. D	Purge and clean the steam wands	Use warm soapy water and a non-abrasive cloth to remove all milk residue.
8. D	Examine the steam wands for cracks or signs of the chrome plating flaking off.	Either condition would require immediate replacement of the wand .
9. D	Remove the drain tray and clean the drain cup	Pour a pitcher of hot water into the drain cup to help rinse accumulated coffee grounds down through the drain hose.
10. A or on-condition	Replace group head shower screen	Even with regular back flushing, the group head shower screens must be replaced periodically .

If we owned hundreds of coffee shops it would be important to look closely at each task and see what is the optimum frequency consistent with high quality at the lowest cost. We might want to abandon the calendar-based system in favor of a utilization (shots) based system. Other factors might include the types of coffee that are popular, weather, water composition, and skill of the operators. It is possible that cutting 10 minutes a day from the PM routine (without any quality sacrifice) in a hundred stores might be worth over 4000 hours a year! The unit based PM for this machine follows (A-Annual, Q-Quarterly, M-Monthly, W-Weekly and D-Daily):

The D-Daily and W-Weekly items might be printed on a placard and mounted directly on the machine. The placard must be mounted where it can be seen. As in factories, the operator should do the daily and weekly items.

Complete training is desirable for success. It is a good reminder that the coffee 'operator' might not have dedicated his or her life to the field and might not have good judgment in this arena. Effective training combined with physical reminders creates a structure that will cause the person to do the right thing (even if the boss is not around).

Generally M-Monthly, Q-Quarterly and A-Annual items are managed in different ways and might be done by specialists or more highly trained individuals. In a busy shop, daily items might become begin or end shift items (done every 8 hours).

String based: Your string-based list is designed to perform one or, at the most, a few short PM tasks on many units at a time. Each machine is strung together like beads on a necklace. Lube routes, infrared scanning, gauge reading and recording, vibration routes, and morning inspection routes are examples of string PM.

String-based PM is particularly important when assets are located near each other or the skill, tools and materials are very specific. When the units are close it might be easier to look at one item on each unit. The inspector's efficiency would be higher because they would be focused on one activity.

Most inspection-only PM's (such as vibration route or infrared survey) are designed as strings. Various types of strings handle almost all predictive maintenance. The reason is that very specific (and usually expensive) tools are needed (that are not normally carried) and specific training is required. Also, when you are in the zone, the inspection route goes faster and more completely.

Advantages: Low training requirements, lower-level mechanic required, job can be engineered with specific tools and parts, route can be optimized, stock room can pull parts for entire string at one time, the procedure lends itself to just-in-time (JIT) delivery of parts or materials, it is easier to set time standards for a string, and the approach offers a good training ground for new people to teach them the plant, allowing new people to get productive quickly. An example of JIT parts delivery might include an air handler filter drop off the first Monday of every month (immediately preceding the filter change string).

Disadvantages: There may be some loss of productivity with extra travel time for several visits to the same machine, a mechanic may not see the big picture (the string person might ignore something wrong outside their string), it is boring to do the same thing over and over, there is no ownership, and it is hard to supervise so that, if a mistake is made (such as wrong lubricant) it is spread to all assets on the route quickly. Another disadvantage is that only a few computer systems (CMMS) support string PM and allow a charge to be spread over several assets.

Current technology has taken some steps to make string PM more responsive. Popular vibration analysis equipment will display a route to each asset and to each point on the asset. In more advanced machines the operator uses a built-in bar code scanner to identify the asset on the route. The advantage is the route order can be changed to suit production or operational conditions.

Future Benefit: This type of task list takes advantage of closely coupled processes. In future benefit PM, you plan PM for the whole train of components whenever a breakdown or changeover idles any essential unit. Some people would argue (perhaps properly) that future benefit PM is not really PM at all because although it can be planned it cannot be scheduled. This author would argue that any maintenance done before breakdown with the effect of reducing breakdown is truly PM.

It could also be argued that future benefit PM is a subset of unit PM, which is true if you only look at one of the pieces of equipment. However, it is untrue when you look at the whole line you are working on with PM personnel and activity.

Future benefit PM is commonly considered in the chemical, petroleum, and other process oriented industries, where processes depend directly on each other. Manufacturing is looking more and more like continuous processes so future benefit PM will become more popular there also. In some areas the crew can accomplish a more complete PM by extending the downtime a few extra minutes. It is usually easier to extend downtime for an hour then it is to get a fresh hour (for PM purposes).

Future benefit PM applies only in specific circumstances. For example, in chapter 3 (on economics of alternatives) we discussed a chemical transfer pump that was connected to a downstream process. In that example, downtime was only charged when the pump was off-line for over an hour (the time it took for the down stream reservoir to empty).

Once the reservoir emptied, the whole line went down. At that point, under future benefit PM, a crew could be dispatched to execute pre-assigned PM task lists on all the equipment forced out of service by the incident. This is an opportunistic approach, but it does give somewhat of an edge compared to waiting for an annual shutdown.

Advantages: Little or no additional downtime, advantage can be taken of existing downtime to PM for a future benefit, the work can become a contest against time, it is easier to manage, and can be exciting.

Disadvantages: The advantage of most PM activity is that it can be planned and scheduled. People can work in a reliable and known environment. Future benefit takes away that benefit of PM and is much more like the thought process in breakdowns. Other problems include not having enough people to do all the work on the list in a timely manner.

Future Benefit PM will also be disruptive to other jobs that were interrupted when the call came in because you cannot predict when your next PM will be needed.

Expenses might increase because you don't have enough crew on-shift and there might be a temptation to call people in for overtime. Finally, future benefit PM can create tremendous mental pressure.

Condition based PM (CBM): This method is a PM mode made immensely more popular by computerized control systems in vehicles, buildings, and factories. Popular SCADA systems, building management, and vehicle computers, can feed data to a condition-based maintenance decision engine.

Condition based maintenance (CBM) presented a problem before the widespread use of the computer for monitoring equipment and building conditions. In some conditions the inspections needed to be done once an hour or more frequently. Certainly a boiler operator could scan the gauges every hour or more, but unless there was someone assigned to the job full time, frequent readings were a significant burden.

There has always been some confusion about whether CBM is PM at all. Using the logic of future benefit PM, CBM is certainly preventive. CBM also resembles preventive maintenance inspections in another way in that it also generates corrective work. Corrective maintenance generated from CBM is plannable and schedulable because usually there is a time lag between the reading and the immediate need for the corrective action.

CBM is the most accurate PM choice for managing critical wear. If the event being monitored deteriorates with critical wear (such as amperage increases while a motor bearing deteriorates) then we have a unique view into the health of the machine.

In CBM, the PM service is based on some reading or measurement going beyond a predetermined limit. The limit can be any measurable event, reading difference (between two readings), or projected trend. CBM is used with statistical process control to monitor and insure quality.

Conditions might include:

- A machine cannot hold a tolerance
- A boiler pressure gets too high
- A low oil light goes on
- Pressure drop across a filter exceeds a limit
- Amp readings have been trending up

Once the condition being monitored goes out of bounds, corrective action is initiated.

Advantages: There is a high probability that some intervention will be needed (fewer false positives), it involves the operator, it brings maintenance closer to production, and it supports quality programs.

Disadvantages: Might act too late to avoid breakdown, usually high skills are needed to design into the system, it can be expensive to implement the first time, it can be planned but cannot be scheduled, and many problems are uncovered that are not maintenance problems.

Example: Condition is that the low oil pressure light turns on, in a Class 8 large truck.

Note that the oil sender circuit is the inspector here.

As with any indication, we have to be disciplined enough to immediately sideline a truck on which the oil light is illuminated. In the trucking field this level of discipline is common because of the consequences of there really being a problem with the oil pressure (loss of the engine). Oil problems are pretty common and usually easy to repair. Without the discipline, all the inspectors and indicators are a waste of resources. One of the advantages of a CBM approach is that the best mechanics can develop, ahead of time, a list of what to do if a specific condition is met.

Example of a corrective action for OIL LIGHT 'ON' condition.

Top off oil if low
Check history file for excessive oil use or recent related work (oil filte change)
Send in sample of oil for analysis
Check for oil leaks
Check tightness of sump plug Verify integrity and presence of plug
Check oil temperature sender
Examine oil cap
Examine cylinders, cylinder seals
Examine oil filter, oil filter seal
Check oil pressure sender
Other specific oil related checks
Correct any problems resulting from above checks.

In the trucking business (as with others), when we have custody and control of the asset we might also perform any other PM that is due at that time on that asset.

Where to get the original PM task list

The task list consists of the items to be done, the inspections, the adjustments, the lube route, and the readings and measurements. The list also includes some indication of the frequency of the task. Sources of task lists are:

OEM (Manufacturer of equipment)
Equipment dealers
Skilled craftspeople experience
Engineering department
Your experience
History, review of your records
Federal, State or local law
Regulatory agencies- EPA, DOT
Third party published shop manuals
Insurance companies
Consultants

There are two types of task and tasklists. **Mandatory**- those tasks required by law, and **Discretionary**-all other tasks.

Thoughts on the OEM task lists

Most people in maintenance enjoy imagining what tasks would be appropriate for an asset. They enjoy using their experience and knowledge for this kind of problem.

They might or might not use the manufacturer's manual. The problem is that there is no linkage between individual tasks and the failure modes that they are to address.

Other people take the manufacturer's task list as an absolute given (which it is, while you are under their warranty). Here are some observations about manufacturer's task lists:

1. Some manufacturers have tremendous knowledge about the failure modes of their equipment based on deep analysis of thousands of units under all types of conditions. You can certainly start with these lists because significant brainpower has been invested in them. Certainly the lists from the large automotive, pump, compressor, mining equipment, or HVAC companies would fall into this category. However, even these lists can be fine-tuned for specific usages, site conditions, or business requirements.

2. Profit drives the task lists of some manufacturers. They want you to over-PM their asset so that they avoid warranty claims.

3. In the OEM business model, a lot of the profit comes from sales of spare parts and service of equipment. Be wary of manufacturers who recommend that you stock large numbers of spare parts. Their spare parts list might include a bunch of items that you are likely never to use (or not use for quite a while).

4. For most manufacturers, ignorance may be the biggest issue. Many small machine manufacturers and some large ones do not use their own equipment. Their engineers might know about the design issues of a pump but they never fixed them or worked with them in service. A big user of the equipment gets to know far more about the equipment than does the manufacturer. You see the equipment every day, you get to fix it, you get to be stuck with the results of your actions, and you learn what it takes to keep the equipment running. Thus, your knowledge base is far more valid.

5. The last issue is that you might use equipment in an unusual service. The manufacturer might be very conscientious, as with members of group 1 above. You are using their equipment "outside the envelope." You might be using it more hours per day, at higher capacities, for different materials, connected to another asset, or under unique controls. I'm reminded of a pick-up truck being used to run a sawmill. The truck was chopped up and welded into the machine. The truck makers could not predict this type of service and consequently you could not rely on their list.

Equipment Dealers

This group can be a significant source of equipment expertise. In certain fields the equipment dealer is the key player and the first line of defense in knowledge about the equipment. An industry where this maxim is particularly true is mobile equipment and related mining equipment.

In this field, the dealer is the primary backup to the in-house maintenance department. In large operations the dealer might have weekly or even daily contact with the maintenance department. Many dealers will supply and modify the OEM task lists and provide training for the maintenance department. Since most dealers in this field also provide service they become outsourcing partners when the in-house department becomes too busy.

Skilled craftspeople and maintenance management experience, Engineering department, History

Once the warranty has expired, it is up to you to modify the tasking to suit your failure experience. The modified task lists will be based on the experience of your skilled trades people and your institutional memory. This is where good records and a well set up CMMS is a great benefit.

Third party published shop manuals

In the heavy truck industries there are a few publishers who have developed expertise in repair and maintenance. These include Chilton, Mitchell's, and a few others. These publishers have developed expertise independent of the OEMs and provide excellent reference books. Their shop manuals include specific assembly/disassembly instructions, time estimates, and tasking lists. Shop manuals of this level of detail are not available in most other industries.

Federal, State or local law, regulatory agencies, such as EPA and DOT (Mandatory PM)

Whenever there is a catastrophe or a highly perceived risk the government will pass a series of laws. These laws are implemented by regulations that are written by agencies created or tasked with enforcement of the laws. Most of the regulations that are detailed to the task list level are in industries that are highly regulated and represent a significant safety or environmental risk to the public.

In highly regulated industries, the regulator may be the primary authority on the tasks to be done. The regulator can also take the role of auditing your task lists to insure that you perform all tasks and that the tasks are complete for the equipment involved. Examples of regulated industries of this type in the US include: FDA-pharmaceuticals, NRC-Nuclear power, FAA-Airlines, DOT-trucking, Joint Commission-hospitals, and some others that primarily inspect for PM related to safety including Dept of Health-Restaurants, ACA-summer camps, etc.

Insurance companies

Several insurance companies have accumulated significant risk-based databases and recommend tasks. In some applications, such as infrared scanning of electrical switchgear and distribution, they even pay part of the bill.

Consultants

Individual consultants and large firms can provide guidance for task lists. Many of these firms are based on the expertise of highly qualified maintenance engineering professionals with long experience in their target industries. In other instances the consultant also provides a CMMS and get their lists from a compilation of users' task lists.

In one example, the consultant was a division of an insurance inspection service. Their data was particularly interesting because they had seen how equipment fails (in the worst possible ways).

PM Frequency: How often do you perform the PM tasks

Determining the frequency of performing tasks is one of the most complex and forceful decisions in all of PM and PdM. If you err on the side of too little frequency you have excessive failure in addition to paying for tasks. If you err on the side of too much frequency then you waste labor and materials every day (and build in a higher cost for your product or service).

There are four ways to determine the correct frequency. Three of these ways will be discussed in this chapter and they include using the manufacturer or other outsider, using failure statistics to predict frequency, and basing frequency on the number of write-ups. The fourth way is the P-F (performance-failure) curve method and will be introduced in Chapter 18.

The first way means that you must take someone else's word for the frequency. Discussed in detail below, this is the route most people choose. It has the advantage of having outside authority behind it and it is always the starting point for all frequency decisions.

Using outsiders is indicated when you have standard equipment in standard service. The bigger companies who manufacturer standard equipment have pretty reliable task frequencies. Using prepackaged lists from outsiders is problematic when you have specialized equipment, 1-off (special) machines or unusual service requirements. In these conditions you might start with the standard lists and use one of the other techniques to modify it.

Simple Statistics (this section partially adapted from the *Handbook of Maintenance Management* by Joel Levitt)

The most common, simplest way (but unfortunately frequently misleading) to determine frequency is to use the MTBF (Mean Time Between Failures) from your history file. This method is particularly useful when you have several pieces of equipment of the same class (like equipment in like service). Increased population of equipment increases the accuracy of the statistics. This method has serious drawbacks because it does not include the issue of how deterioration takes place. This issue is covered in a more complete discussion of frequency in Chapter 18 called the P-F curve (Performance-Failure). Most deep PM analysis techniques use P-F methods for PM frequency.

Statistics are a powerful set of tools that can help improve the PM system. The simplest idea is the mean time between failures (MTBF). For our purposes the MTBF is the same as the average elapsed time or utilization (mileage, machine hours, even tonnage in a mine) between failures. To calculate the mean, add up all the elapsed time (or tons, bottles, KWH or whatever measure of equipment use is adopted) between failures and simply divide by the number of readings.

The MTBF is used with the standard deviation (SD). The SD measures the variability of the measurements. For example, the mean of the three readings (1, 10, 250) is 87. The mean of a second group of three readings (79, 89, 93) is also 87. As you can see the mean doesn't express a sense of the variability of the readings. The SD of the first distribution is 115 while the SD of the second is 5.9. As the SD gets smaller the predictive power of statistics for the purpose of PM frequency assignment improves.

How to calculate the standard deviation (SD)

1. Calculate the mean (total of all readings/number of readings)
2. Subtract each reading from the mean: Difference = Mean - Reading
3. Multiply each difference by itself, SQ difference= (difference)2
4. Add the SQ difference for each reading Sum= Sum of all SQ difference
5. Divide the Sum by the number of readings Variance = (Sum)/# readings
6. The SD is the Square root of the Variance: SD = Sq root of (Variance)

The Normal Curve represents ideal reality where there is one major failure mode

The normal (or bell-shaped) curve is a graphic representation of a large number of failures where one failure mode dominates (the more one dominates and the more readings the smoother the curve).

You can use the standard deviations (SD) just calculated to divide a normal distribution into partitions that are extremely useful to help you choose a PM frequency. The size of the partition is called one standard deviation (SD). The useful property of the SD is that 68.27% of your readings will be within 1 SD of the MTBF. (That is $\pm$1 SD) and 95.45% of your readings will be within 2 SD of the MTBF. ($\pm$2SD).

You choose your inspection frequency based on your desired failure rate. 1 SD gives you a certain cost and a certain amount of failure, 2 SD gives you a greater cost and a lower failure rate.

Using MBTF for failure analysis

Failure analysis reviews the failures and using statistics comes to some conclusions about their frequency. The technique of failure analysis is to determine the elapsed utilization between incidents of failure (MTBF) and the time it takes to put the asset back in service (MTTR).

Detailed failure analysis that is statistically valid is not for the faint hearted. Accuracy dictates large populations of failures. Good engineering practices dictate tracking each mode of failure separately. These details generate enormous amounts of data and take significant resources. Some CMMS have primitive statistical capabilities. The best CMMS will collect the data needed for statistical analysis and export it to a spreadsheet or specialized statistical package.

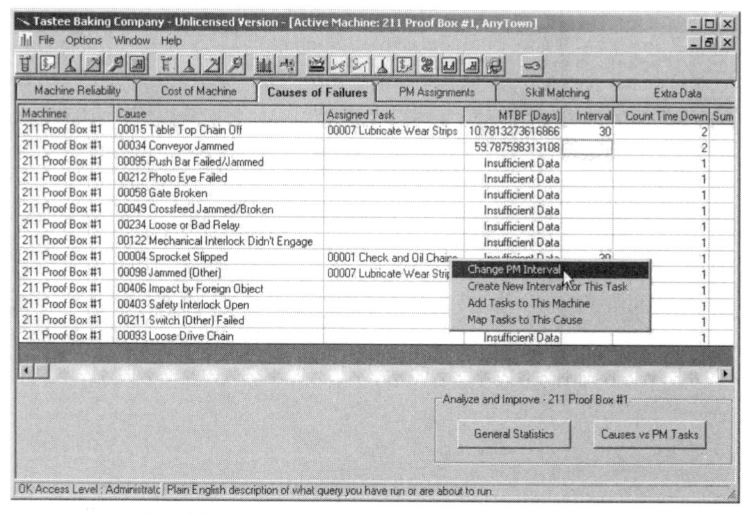

One of many screens from Maintsmart designed to help the PM designer. This is a CMMS with powerful analysis capability for PM (see resources section)

Failure experience (both frequency and severity) feeds back into task list

Failure analysis can be a major tool in the establishment and updating of PM systems. If failures were too frequent (in relation to the frequency of the PMs) then increases in the depth of the task list or in the inspection frequency would be required. If failures dip too low then the reverse may be true and too much money is being spent on the PM activity for that component. Note that this is an economic analysis. Task list items that are directly concerned with life safety are not included in this analysis.

Use the information to setup a PCR program (discussed in more depth in Chapter 18). If the failure of the component follows a normal distribution then PCR

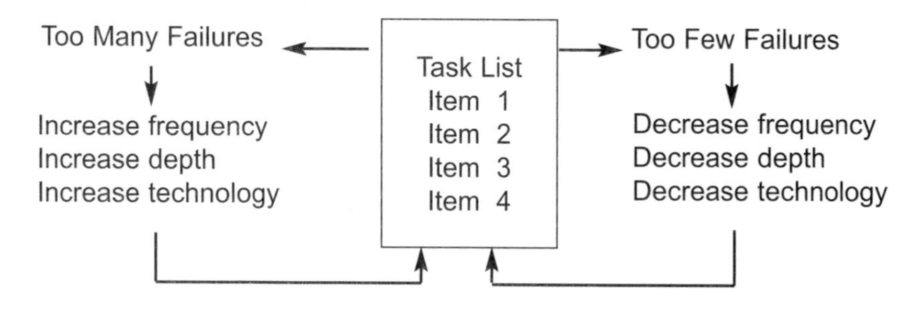

Failure history impacts task list through feedback

might help. Planned component replacement frequencies should be chosen to program an allowable failure rate. If you have 1000 failures per year under current conditions and change out the component at 1 SD before the mean time between failures (MTBF) then the new failure rate will be 15.9% of the old rate. In the new scenario you will have 159 failures next year. The aircraft industry uses 3 SD (3 sigma's), which gives them less than 1% of the original failure rate.

Other uses of MTBF statistics

Use the information to compare two makes of components. You might want to compare two makes of bearings to choose one over the other for a particular application. Look at the MTBF for each component and factor in the cost

Failure analysis can be used to interpret the results of experiments and provide data for efficient decision-making. An example is looking at compressor failures for synthetic versus natural oils.

Maintenance departments constantly evaluate their specifications. Failure analysis can help improve specifications. Although there are many factors to the choice of a component or system (like price), MTBF and MTTR (mean time to repair) should be among them.

Another way to set frequency: Look at the frequency corrective actions

We expect a certain number of observable problems per hundred or thousand inspections. We can infer the proper inspection frequency by observing the number of write-ups. Some organizations use the standard that if they don't get a reportable item every other PM then they are inspecting too frequently. Don Nyman uses the standard of 6:1 in his course text *Maintenance Management.*

The task list should be designed to capture information about or direct the attention of the inspector toward critical wear areas and locations. If you are inspecting an expensive component system, many inspections might go by without any reportable changes. Depending on the economics, you may want to continue to inspect in order to capture the change when it happens.

Always continue to inspect life safety systems. An OSHA mandated inspection of an overhead crane hook might have a ratio of inspections to observed deficiencies of 10,000:1 or greater. Do not include statute driven inspections (boilers, sprinklers, etc.)

Sources of Frequencies

The two core questions on micro-PM are what tasks and how often. As discussed above there are many sources for task lists and frequencies. The first source for inspection frequency is the manufacturer's manual or recommendation. Ignoring it might jeopardize your warranty.

The manufacturer assumptions for how the machine is being used might be different from your usage. For example one manual recommended a monthly inspection for a machine. When the manufacturer was questioned it came out that the assumption was made based on single shift use. The factory used the machine around the clock and was getting excessive failures even with recommended PM frequency. The solution was to increase the frequency to weekly and the failure rate dropped to a reasonable level.

Some manufacturer's maintenance manuals are concerned with protection of the manufacturer and limiting warranty losses. Following that manufacturer's guidelines may mean you will be over-inspecting and over doing the PM needed to preserve the equipment.

The law drives certain inspections (in the USA agencies that require inspections include EPA, OSHA, State, and DOT). You have a certain amount of flexibility in the timing of these inspections. Consider scheduling them when another PM is also due. While you have the unit under your control, you also perform the in-depth PM to improve efficiency.

Your own history and experience are excellent guides because they include factors for the service that your equipment sees, the experience of your operators, and the level and quality of your maintenance effort.

For almost any measure to be effective the PM parameter (such as cycles, days, etc.) must be driven from the unit level (unique PM frequency table for each unit), or from the class level (a class of equipment is defined as like units in like service), and have the same PM frequency. For example a pick-up truck would have a very different frequency than a dump truck even though they are both trucks.

The PM inspection routines are designed to detect the critical wear point and defer it into the future as much as possible. Since we cannot always see the wear directly, the goal is to find a measure that is easy to use and **is directly proportional to wear.** Traditionally two measures have been used: utilization (cycles, tons, miles, hours), and calendar days. These measures are called "clocks." Other measures mentioned below are not only possible, but in some cases more accurate.

Days or calendar based: In this, the most common method, the PM system is driven from a calendar. (Example- every day, grease the main bearing, every 30 days replace the filter, etc.)

Advantages: Easiest to schedule, easiest to understand, best for equipment in regular use

Disadvantages: PM might not reflect how the unit wears out; units might run different hours and require different PM cycles (example, one compressor might run 10 hours a week and another might run 100 hours)

Meter readings: (example- change the belts after the compressor runs 5000 hours). This is one of the most effective methods for equipment used irregularly.

Advantages: relates well to wear, is usually easy to understand

Disadvantages: Extra step of collecting readings, hard to schedule in advance unless you can predict meter readings,

Use: Second most common method. The PM system is initiated from usage such as: perform the PM after every 50,000 cases of beverage, or overhaul the engine every 10,000 hours or 500,000 miles. Some theme parks even use the number of guests passing through the turnstiles. Building management systems and SCADA systems track hours for components.

Advantages: Utilization numbers are commonly known (how many cases we shipped today). The parameter will be well understood, should be very proportional to wear, not hard to schedule after the production schedule is known, but may be harder to predict labor requirements in a future month or year. The production number might be obtainable from another system.

Disadvantages: the information system might not accept this type of input, and extra labor may be needed to take readings or collect data.

Energy: The PM is initiated when the machine or system consumes a predetermined amount of electricity or fuel. The asset would have a meter or some other method of directly reading energy usage. This method provides an excellent indirect measure of the wear situation inside the device and of the overall utilization of the unit. You probably are already collecting some energy data for other reasons. Energy consumption includes the variability of rough service, operator abuse, and component wear (increased friction). The method is used extensively on boilers, construction equipment, and marine engines, and data collected can be used for other purposes (such as increasing equipment efficiency.

Advantages: Very accurate measure of use in some equipment raises consciousness about energy usage.

Disadvantages: Need to wire watt meters or oil meters into all equipment to be monitored, hard to schedule ahead of time without a good history, extra labor is needed to take readings or collect data.

Consumables (example: Add-oil: The amounts of additions to hydraulic, lubricating, or motor oil are tracked.) When the added consumable exceeds a predetermined parameter then the unit is put on the inspection list. This method provides a direct measure of the situation inside the engine, hydraulic system, gear train, etc. Wear and condition of seals are directly related to lubricant consumption.

Advantages: Will alert you if there is a leak,

Disadvantages: Very specialized, very hard to schedule in advance, hard to collect data

On condition measures (such as Quality): The PM in this case is generated from the inability of the asset to hold a tolerance or have consistent output.) It could also be generated from an abnormal reading or measurement. For example, a low oil light on a generator might initiate a special PM.

Advantages: Responds well to customer needs

Disadvantages: Almost impossible to schedule, cause is frequently not in the maintenance domain, might be too late.

PM Levels and resetting the clock

PM lists are hierarchical. In general, lower frequency tasks have more depth and are more time consuming than more frequent tasks. Some systems are designed so that the less frequent tasks can get out of sync with the more frequent ones. When you start a system off, the weekly, monthly, quarterly, and annual tasks are in sync. That means, for example, the quarterly task coincides with the third iteration of the monthly task.

The system behind PM scheduling can be set up in a variety of ways. In some systems the tasks exist independently of each other. This arrangement allows maximal flexibility but does introduce the possibility of confusion and duplication of effort. In an independent system, doing a quarterly task has no impact on the scheduling or execution of the monthly routine. Consequently, in an independent system the tasks can get out of sync.

For example on Wednesday the Annual PM instructs the mechanic to replace the oil and the following Tuesday the Quarterly PM instructs the mechanic to top off the oil. To fix this problem system designers have introduced two extras. One forces a reset to zero of the clocks for "lower" PMs. The annual task would (properly) reset the clock to start over from zero on the quarterly top-off task. If your system has this capability it is important to be sure that the task list for the longer task includes the task list for the shorter tasks.

The second extra is a 'look ahead' function. When a PM is generated, the system looks ahead for any other services on that equipment. It notifies the manager, supervisor, scheduler, or mechanic that additional services are due soon. Often the two services can be combined for greater efficiency.

Alternate designs don't have this problem (but are much less flexible) because an annual inspection is actually a twelfth monthly PM. The Annual PM comes along every twelve months.

Loss of synchronization of PM schedule

Another issue is what to do with scheduling PMs if they are delayed or just not done. Generally we would pick #2 below for both. Regulatory requirements may require option 1 in certain fields.

Case 1

A monthly PM for January 1 is done January 17:

1.Schedule next PM for February 1

2. Schedule next PM for February 17. Report on slippage.

Case 2

A monthly PM for January 1 is not done at all:

1. Schedule next one for February 1 and have two outstanding
2. Wait for the January PM ticket to be closed out before generating any more PMs and show the January PM as increasingly overdue.

Of course the problem of choosing #2 in each instance is slippage. All the effort making sure that your PM loading for each month and week is balanced and within labor hour guidelines, flies out of the window after a few months. If you use strategy #2 you might consider reloading and resetting the PM schedule every year or two to realign the hours.

What do you do if you find yourself with PMs that never seem to get done? Can you question if that PM belongs on the PM list at all? Check your commitment to the PM on this particular asset. Consider convening the PM taskforce and trying to either eliminate the asset from the list, cut back on frequency, cut back on tasks, re-engineer to eliminate or reduce the need for PM, or re-engineer the PM tasks to make them capable of being done quicker, more easily and more conveniently.

Task List Development

The goal of task development is to end up with a task list where all tasks serve a purpose and are cost effective, there are no duplicated tasks, and all economically preventable failures are preventabled. The second aspect is that all task lists should be designed, where appropriate and desirable, to use higher levels of computerization of monitoring.

More than the general population maintenance people tend to be risk averse. This attitude probably comes from a professional life dealing with life threatening or production stopping machines, materials, and tools. Whatever the reason, many task lists designed by maintenance professionals without analysis usually err toward excessive PM. The professional's tendency is to add tasks and increase frequency. If you or your staff went to school at the University of Hard Knocks, analysis is essential!

Task Analysis

Tasks, once approved, will consume money from that point in time forward. Organizations should be very careful which tasks they approve because of the permanent added costs they represent. The second area of concern is task focus. Each task developed should be focused on at least one probable failure mode.

It is important to reiterate that the task list represents the **accumulated knowledge** of the manufacturer, skilled mechanics, engineers, and managers in the probable failure modes. In addition to probable failure modes the list represents strategies for detection and correction of failures.

When designing PM Task Lists, look for the <u>most likely</u>, the <u>most expensive,</u> or the <u>most dangerous</u> types of failures.

Definition: Class of equipment (there are many other names such as equipment category or type). Class is defined as **like equipment in like service**. A trash truck on a pick-up route in Baltimore could be in the same class as a trash truck in similar service in Chicago. But the same truck in line haul service (taking the trash from a transfer station 50 miles to a landfill) would be in a different service and different class.

For PM task generation purposes a group of units in one class should be aggregated. The larger the class, the more accurate are any statistics and the more complete are failure mode tables.

Process

1. Generate an equipment history from your CMMS or from your manual records. If you have similar assets in similar kinds of service, do this exercise for all

units in that class at the same time. In keeping with the three-part goal listed above, look at the most dangerous, costly, and frequent failures.

2. List all the failures on a chart with the consequences. If possible, quantify the consequences.

3. Chart out MTBF (remember the more units in the same class, the merrier) by class and by component system.

4. Once this base data is collected call a meeting for the purpose of design (or redesign) of a particular task list.

5. Potential members would be maintenance workers (with long memories with that asset), machine operators, planners, supervisors, representatives from the manufacturers, representatives from the equipment dealer, and PM facilitators.

6. The meeting should review any hard data from the CMMS and then proceed.

Be sure to look for hidden protective devices

Buried in many devices and machines are devices designed to protect us from harm or to protect the machine from a small problem that might grow to destroy the entire machine. Hidden protective devices include fuses. A fuse is designed to fail and to interrupt the flow of electricity to prevent a fire. Other devices, such as a low oil light, are designed primarily to protect the equipment. The challenge of maintenance of these devices is to verify that they function and that they work to specification.

It is essential to develop tasking to look after these protective devices. Many devices are removed and discarded on a periodic basis (planned discard). In other

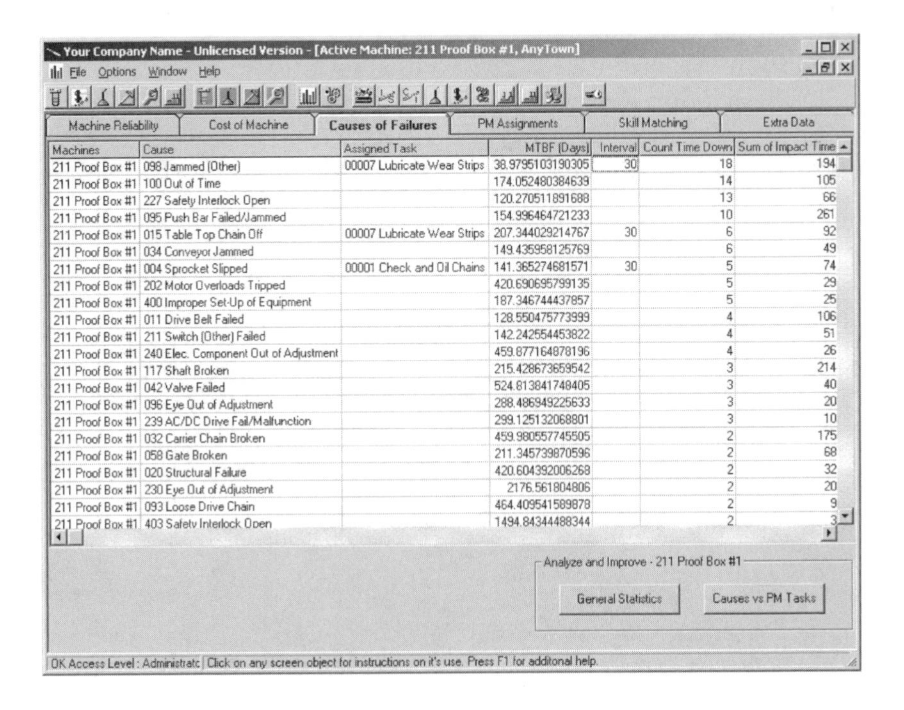

Machines	Cause	Assigned Task	MTBF (Days)	Interval	Count Time Down	Sum of Impact Time
211 Proof Box #1	098 Jammed (Other)	00007 Lubricate Wear Strips	38.9795103190305	30	18	194
211 Proof Box #1	100 Out of Time		174.052480384639		14	105
211 Proof Box #1	227 Safety Interlock Open		120.270511891688		13	66
211 Proof Box #1	095 Push Bar Failed/Jammed		154.996464721233		10	261
211 Proof Box #1	015 Table Top Chain Off	00007 Lubricate Wear Strips	207.344029214767	30	6	92
211 Proof Box #1	034 Conveyor Jammed		149.435958125769		6	49
211 Proof Box #1	004 Sprocket Slipped	00001 Check and Oil Chains	141.365274681571	30	5	74
211 Proof Box #1	202 Motor Overloads Tripped		420.690695799135		5	29
211 Proof Box #1	400 Improper Set-Up of Equipment		187.346744437857		5	25
211 Proof Box #1	011 Drive Belt Failed		128.550475773999		4	106
211 Proof Box #1	211 Switch (Other) Failed		142.242554453822		4	51
211 Proof Box #1	240 Elec. Component Out of Adjustment		459.877164878196		4	26
211 Proof Box #1	117 Shaft Broken		215.428673659542		3	214
211 Proof Box #1	042 Valve Failed		524.813841748405		3	40
211 Proof Box #1	096 Eye Out of Adjustment		288.486949225633		3	20
211 Proof Box #1	239 AC/DC Drive Fail/Malfunction		299.125132068801		3	10
211 Proof Box #1	032 Carrier Chain Broken		459.980557745505		2	175
211 Proof Box #1	058 Gate Broken		211.345739870596		2	68
211 Proof Box #1	020 Structural Failure		420.604392006268		2	32
211 Proof Box #1	230 Eye Out of Adjustment		2176.561804806		2	20
211 Proof Box #1	093 Loose Drive Chain		464.409541589878		2	9
211 Proof Box #1	403 Safety Interlock Open		1494.84344488344		2	3

devices a test of the protective device is in order (running water out of a boiler until the low water cut-off opens). In any event, development of procedures to verify both the operation and specification is important.

Task design always starts from failure modes.Thorough understanding of exactly how your equipment fails in your operating environment is essential. Sometimes your CMMS can be a big help. In it are all failures, dates, and (hopefully) existing PM task lists. One system that is a stand out in this regard is Maintsmart (see resources section).

The screen on the prior page shows all the failures for a Proof box are shown with MTBF. You can go through and assign tasks and choose intervals.

Basics in task generation (once the process is started for an asset)

Step 1 Define the failure modes you want to attack

Description of Failure avoided
What component (sub-component) is being looked at?
Cost and consequences of failure avoided (if a safety or environ-mental catastrophe is involved, look to redesign)
Estimated PF interval for failure
Estimated annual frequency of failure Annual cost for failure

Step 2. After examining the asset, drawings, and history, choose a proposed task

Complete description of task
What failure mode is being addressed (from above) and how.
What clock is best for this type of task (days, utilization, energy,condition,other)?
Proposed frequency of task, why this frequency?
Task time Materials required
Task cost Annual task cost

Step 3 Analysis of the details

If necessary, complete simplified drawings to show how the task is done. Are there lock out, tag out, confined space entry, or other safety issues?
Are there environmental consequences to the task such as possible spillage or release of gases?
Develop specifications and recommendations for the task.
Recommend type of task list (unit, string, future, etc.)
What skill level is required for the task?
Is a special license needed for the task?
Is there a legal liability issue?
Can this task be done by the operator or can the task be in-sourced elsewhere?
Is a contractor a better choice for this task?
Will doing this task impact any other task (changing oil impacts oil analysis)?
Estimate the time between detection and failure if this is an inspection
Number of components that this task is addressed to
How long will the task take if you are already at the unit with the tools on a cart?
Special tools required Is this task seasonal?
Who will have to be notified?

TLC
Tighten, Lubricate, Clean

TLC (Tighten, lubricate, clean)

TLC means Tender Loving Care. When we apply TLC to machinery we get: tighten, lubricate, clean. Keeping equipment trim and clean will extend the life and reduce the level of unscheduled interruptions. This approach or strategy is appropriate for all maintenance departments, even those with no support from top management or maintenance customers. **TLC is the simplest way to reduce breakdowns.**

The climate seems to be against TLC. As firms experience downsizing and de-staffing one of the first services to go is TLC. When we read the latest trade journals and listen to the latest papers at conferences we hear and read that time based (or interval based) PM is obsolete. At a recent AFE (Association of Facility Engineering) annual meeting there were 35 papers or sessions presented and none of them spoke about improved TLC.

Yet studies find again and again that dirt, looseness and lack of proper lubrication cause the bulk of equipment failures. TLC is the core of TPM's increased reliability. Examples here are partially adapted from TPM Development Program by Nakajima.

One company found that 60% of its breakdowns were traceable to faulty bolting (missing fasteners, loose or misapplied bolts)

Another firm examined all its bolts and nuts and found 1091 out of 2273 were loose, missing, or otherwise defective.

The JIPE (Japanese Institute of Plant Engineering) commissioned a study that showed 53% of failures in equipment could be traced back to dirt, contamination, or bolting problems.

Effective TLC can impact other costs. One firm reduced electric usage by 5% through effective lubrication control

Cleaning

Dirty equipment creates a negative attitude that adversely impacts overall care. Inspectors cannot see problems developing and mechanics don't want to work with the equipment. Dirt can increase friction and heat, contaminate products, cause looseness from excessive wear, degrade the physical environment, cause potentially lethal electrical faults, contaminate whole processes (as in clean rooms), and demoralize the operator.

Cleaning is a hands-on activity. Someone who cleans a machine with their eyes open will see all sorts of minor problems and ask themselves questions about how the equipment works and why it is designed the way it is. This hands-on approach is the key to what is called TPM. TPM (Total Productive Maintenance) makes the operator a key player in the PM program.

This hand-on approach will also increase the person's respect for the machine. This process of cleaning, seeing, touching, and respecting the machine is essential to increase reliability. As a result of the questions and observations made by people doing the cleaning, the operation and maintenance of the machine can be improved.

Quick idea

The people cleaning the machines have the best chance of detecting any failure. As they touch and look at the machine, loose bolts should shout to them.

Part of the cleaning process is looking for ways to make cleaning easier or maintenance avoidable. Perhaps the source of dirt should be isolated to reduce the need for cleaning. In other areas the machine should be moved or rotated to facilitate access. The TPM Development book lists seven steps to a cleaning program:

Cleaning Program Checklist

1. Cleaning main body of machine, checking and tightening bolts.
2. Cleaning ancillary equipment, checking and tightening bolts
3. Cleaning lubrication areas before performing lubrication
4. Cleaning around equipment
5. Treating the causes of dirt, dust, leaks, and contamination.
6. Improving access to hard to reach areas.
7. Developing cleaning standards

Keep area clean

Keeping it clean is not only a PM issue. Cleanliness also promotes safety and positive morale. Cleanliness is important in rebuilds, in major repairs, and even in small repairs. Any mechanic in the business for a length of time can remember a perfect repair gone badly because of dirt.

With all of the attention being paid to dirt and cleaning one would imagine organizations would take extra steps to exclude dirt when they do major repairs. How many professional maintenance organizations take control of the physical environment with work tents, plastic drapes, or other measures to exclude dirt and contamination (when there isn't an environmental issue).

Keeping the maintenance shop itself clean should be a goal of the maintenance program. Issue a periodic work order to clean up the shop. Also look at eliminating the sources of dirt and clutter such as misplaced trash containers, lack of proper storage, broken tools, bad ventilation, inadequate lighting, and benches that are too small.

Bolting

"Bolts are tightened by applying torque to the head or nut, which causes the bolt to stretch", (refer to Machinery's Handbook). In interviews with people on the shop floor responsible for mechanical maintenance I was surprised to learn that most people don't know the basics of bolting. It would be useful to get people trained (your old timers might have practical knowledge and not know the basic engineering of bolting).

Misconceptions

Using a torque wrench is infallible: Not always, because of friction. Remember the goal is to stretch the bolt. This stretching clamps the joint. If there is rust or dirt, greater torque will be needed to achieve the same level of elongation. If there is grease, the torque required will be greater to achieve the same elongation.

It doesn't matter what the joint looks like when you pick a torque setting. Different joints require different amounts of torque. A joint in tension requires a different torque setting than a joint in shear. A joint in compression has significantly less torque requirements than either of the others.

All bolts of a particular size should be torqued to the same degree. Bolts come in grades. The range of strength between a grade 1 and a grade 8 bolt is almost 4 to 1. That means the the torque needed to stretch the bolt could vary as much (depending on the application) as well.

Once you properly torque the bolt you're done. It is a well-known problem in mobile equipment that bolts loosen up in the first 500 miles or 25 hours and should be re-torqued. This loosening results from high points on the nut, inside the bolt head, or in the work being bolted, dirt caught in the joint, and the bolt head (or the bolthole) being out of square. After some vibration and temperature cycles the friction problem is resolved but now the bolt is loose.

No problem with a missing bolt if there are others intact. Loose or missing bolts are a major source of breakdown. Even a single missing or loose bolt might cause a failure. While properly engineered joints are designed with structural redundancy, each fastener is important. The bolt is tightened and it stretches. This elastic stretch creates a clamping force to engage friction between the pieces being bolted. The number and spacing of the bolts spreads the clamping force evenly over the joint. A single missing bolt can reduce the clamping force locally, which impacts friction. This friction is essential to prevent unwanted motion and vibration.

In most assemblies, the looseness contributes to vibration that in turn increases looseness. In electrical joints connected by the pressure of a bolt, looseness is usually the result of thermal expansion and contraction. The space that looseness creates promotes oxidation, increasing resistance that expands and contracts the joint, and that causes more looseness. In other words, loose bolts beget loose bolts.

Ouch the $10 million nut

The misapplication of one nut cost an air charter company $9,600,000. In 1990 there was a plane crash in the Grand Canyon. The nut holding the propeller on a small tour plane came loose causing the propeller to fly off. The jury awarded $9,600,000 for negligence in maintenance practices.The tour operator had a $10,000,000 deductible insurance policy. If a main nut holding a propeller can be missed, what is the chance that you have nuts working their way loose right now as you read this section?

Idea for action: The easiest technique is to scribe a line on the nut and on to the machine frame when the nut is tightened correctly. This scribed line will stay intact (a single line) as long as the nut doesn't move.

When equipment is engineered, the rules of good bolting should have been followed. Much of the process of maintenance is correcting mistakes or deviations from good engineering practices. Many rules concern the size, pattern, torque, and type of fastener. Other rules include head location (nut is accessible), use of lock washers, use of flat washers, and bolt length.

In most facilities there are no well-known standards for tightness for task lists with tasks like "check base bolts and tighten if loose." Are there standards of this type in your organization? Are they followed and understood by the workers tasked with bolting?

Bringing equipment to specification is sometimes a lengthy job. As mentioned, fleet vehicles are brought in after 1000 miles to tighten everything up. The short run-in period gives the bolts a chance to set. This same strategy is **not** well followed after factory rebuilds or when work is done in buildings.

Good bolting practice takes a while to teach and is not necessarily intuitive.

Lubrication

Lubrication is the Rodney Dangerfield of the maintenance field. It gets no respect. It is assumed by people peripherally associated with maintenance that anyone who can find a zerk fitting and squeeze a handle can be a lubricator. Maintenance experts know that tribology is a field in which you can get a PHD. They also know that a good person in the lubricator's role can save a plant, building, or fleet, thousands of dollars in breakdowns and potentially millions in downtime and accident prevention.

The University of Leeds in the UK has one of the more active Tribology departments. To give you an idea of what that entails, their current projects include research into lubrication for the following components or areas:

1. Piston rings and piston assemblies
2. Cams and followers
3. Engine bearings
4. Engine friction modeling
5. Engine components including belts and pumps.
6. Thermal elastohydrodynamic lubrication
7. Non-Newtonian lubricant effects coupled with roughness influences
8. Elastohydrodynamic lubrication in continuously variable ratio transmissions
9. Soft materials such as rubbers in seals and auricular cartilage with an emphasis on asperity deformation effects

Failures to lubricate are always the result of several factors. A leading factor is poorly designed or installed equipment where the points are too hard to get to or there are just too many points. Other factors include too many different lubricants used, not enough time allowed, lack of standards, and a lack of motivation of the worker. The lack of motivation can usually be traced back to a lack of knowledge of the importance of lubrication to reliability, poor self-esteem resulting from the job being a bottom of the barrel type job, and a lack of training and feedback of how the job is done.

Entry-level operators take weeks to learn basic lubrication. In Japan, the standard course for TPM technicians is 12 weeks. With a few exceptions, mechanics in western plants rarely have any formal training in lubrication. This loss is reflected in the high number of lubrication breakdowns.

We assume that journeymen mechanics are experts in lubrication. Frequently they know only what they've seen and tried. This might be only a small subset of the possibilities and might also be wrong. If you examine the Machine Lubrication Technician job description published by the Lubrication Council (see resources) you will see several important skills.

Level I Machine Lubrication Technician Job description

The purpose for the Level I Machine Lubrication Technician (MLT) certification is to verify that technicians practicing in the field of machinery lubrication, as it is applied to machinery condition monitoring and maintenance, are qualified to perform the following tasks:

- Manage lubricant delivery, storage and dispensation.
- Manage a route for machinery re-lubrication and/or inspection.
  Properly change and/or top up the oil in mechanical equipment found in common industrial sites.
- Use simple techniques to select lubricants with the proper base oil and additive system for machinery commonly found in industrial settings.
- Use simple techniques to select grease lubricants appropriate for machines commonly found in industrial settings.
- Use simple techniques to select grease application methods (including automated delivery) that are least intrusive and most effective for machines commonly found in industrial settings.
- Use simple techniques to estimate re-grease volume and intervals for machines commonly operated in industrial settings.
- Properly maintain automatic lubrication systems (auto-grease, mist systems,etc.)
- Employ basic oil analysis techniques to identify and troubleshoot abnormal lubricant degradation conditions, and use simple techniques to adjust the lubricant specification accordingly.
- Common job titles for the individual who would become Level I MLT Certified include Lubrication Technician, PM Technician, Millwright, Mechanic, etc. Generally, this individual has regular contact with the machine and has routine

influence over the condition of lubricants and hydraulic fluids in use. The individual is likely to be directly involved in the machine lubrication process.

Tests and certifications in Lubrication

In the US there are tests, training, and certification for lubrication expertise. Maintenance Technology Magazine (see resource section) has compiled a list of courses, tests and certifications. Three tests are offered by the International Council for machinery Lubrication (see resource section):

Machine Lubricant Analyst I
Machine Lubricant Analyst II
Machine Lubricant Technician I

Each of these levels requires a 3 hour, 100-question multiple-choice test and is good for 3 years. The different levels cover both predictive analysis capabilities and standard lubrication training.

The second group is a series of tests for learning the science of lubrication itself. Sponsored by the Society of Tribologists and Lubrication Engineers (see resources) they are 150 question tests and their certification is also good for 3-years.

Oil Monitoring Analyst I
Oil Monitoring Analyst II
Certified Lubrication Specialist

Mistakes

Mistakes in lubrication can be devastating. Unlike some other maintenance practices a mistaken lubricant could be spread to all machines in an area in one afternoon as the lubrication route is completed. We occasionally hear horror stories of people substituting the wrong lubricant (in some reports, the substitute was not even a lubricant!)

Case Study: A very expensive lubrication mistake almost caused millions of dollars worth of damage on the drawbridges that cross the St. Lawrence Seaway. These 35-year-old drawbridges (at the time) were activated by 2 cables, which rode on 35-foot diameter pulleys, mounted on steel shafts. Partial cracks were found forming in the shafts. The engineers determined that the cracks were caused by corrosion. A review of the PM work orders for the last 20 years showed that the lubrication was being done at the specified frequency with the correct lubricant.

A Tribologist was brought in to review the whole application. He found that the original drawings and specifications called for a lubricant that was inappropriate for a marine environment. The problem took 35 years to manifest itself. Ask yourself this, if an engineer and the people who checked the drawings made a mistake about the functional qualities of a lubricant what is the probability that the lubricants you've been using are still the best ones today?

Too many choices lead to problems

One issue is that many plants use too many different lubricants. In some applications you can standardize on the 'better' product and save money through buying larger quantities. The cost of the lubricant itself is usually the smallest element of the whole picture. If changes are made to the lubricant specifications, it is important to document them (include your logic for the change). In most facilities the lubricants were chosen a long time ago and the reasoning is lost in time.

For lubrication to be successful, the people involved need to understand why they are doing the lubrication, how to do it, where to do it, and with what. Drawings, charts, diagrams, and photographs (with appropriate legends) are useful in the process. The lubricator must also understand the implications of excessive lubrication.

One of the biggest areas where cleaning and lubrication overlap is in the cleaning and examination of the lubrication points. Clogged, dirty, or broken lubrication fittings compromise the whole effort. Initial cleaning should highlight these issues and correct them.

Lubrication Check List (partially adapted from TPM Development Program)

1. Are lubricant containers always capped?
2. Are the same containers used for the same lubricants every time, are they properly labeled?
3. Is the lubrication storage area clean?
4. Are adequate stocks maintained?
5. Is the stock area adequate in size, lighting, and handling equipment for the amount stored?
6. Is there an excellent long-term relationship with the lubricant vendor?
7. Does the vendor's sales force know enough about tribology to solve problems and do they periodically tour the facility and make suggestions.
8. Is there an adequate specification for frequency and amount of lubricant?
9. Are there pictures on all equipment to show how, with what and where to lubricate and clean?
10. Are all zerk fittings, cups, and reservoirs, filled, clean, and in good working order?
11. Are all automated lubrication systems in good working order right now?
12. Are all automated lubrication systems on PM task lists for cleaning, refilling, and inspection?
13. Do you have evidence that the lubrication frequency and quantity is correct as specified (oil film on moving parts, freedom from excess lubricant)?
14. Is oil analysis used where appropriate?

Save Money by rethinking

Consider eliminating time-based oil changes in large equipment. An oil change can cost $1500 or more and might be unnecessary. Under normal operating conditions there are three reasons to change oil: contamination by dirt, water and metals, changes in the additive package for corrosion resistance or cleaning, and changes to the properties of the oil such as viscosity.

The strategy is to use oil analysis to see if the oil is still in good shape. Problems in any of the three areas can be detected by oil analysis. Thus, the oil change will be based on the condition of the oil as determined by analysis.

The second part of the equation is to either mount a bypass low-micron filter on the equipment (that continuously cleans a small percentage of the oil very well) or purchase a filter cart and periodically (and thoroughly) clean the oil in place.

If the oil is kept clean it will last 3-5 times longer. By performing oil analysis you add the advantage of a predictive inspection that will alert you when abnormal wear is taking place. This overall approach will result in lower overall costs and higher reliability.

Automated Lubrication Equipment

One way to improve the lubrication program is the judicious use of automation. There have been significant improvements in the reliability of automatic lubrication systems. These systems can now inexpensively be retrofitted to existing equipment on a one or multiple point basis. They provide a level of repeatability and reliability unmatched by most manual systems.

In Maintenance Management by the late Jay Butler, the advantages of automated lubrication are listed and they include reduction in the number of people needed to perform the lubrication, improvement in the amount of lubrication dispensed, reduction in the amount of contamination, insurance against missed cycles due to sickness or reassignment, reduction in the number of interruptions to use of the equipment. The end result is lower downtime, reduced breakdowns, and reduced cost of operation.

In the transportation field, lubrication is critical. A seized 's' cam or slack adjuster in a trailer axle can fail either actuated or un-actuated. When it hangs in the actuated position the driver can lose control of the rig causing jack-knifing and a potential accident. Since the early 1980's, Lubriquip has been providing single-point (semi-automatic) systems. In these systems every lube point is piped to a central location. The mechanic uses the grease gun at the central point. This semi-automated mode saves 25 minutes per trailer per month. Other savings include reduced contamination, reduced missed points, and savings in lubricant. The system costs about $250 per trailer. The system will report if a point is clogged and will count the number of lubrication cycles performed.

The biggest mistake in the use of automated equipment is that organizations forget to add the automated lubrication equipment to the PM task list. These systems have to be filled, inspected, repaired, and cared for (TLC).

Case study:

Before lubricant automation, a machine had five to ten lubricant-related bearing failures a year. It now experiences none. This record translates into 30 to 60 hours of additional machine time and profit gains of $100,000 annually. The plant reports a decrease in total maintenance downtime from 470 hours before lubricant automation to 140 hours a year after the implementation of automatic lubrication on one major machine. Grease consumption is now down to 85% of the amount used previously with manual lubrication.

According to Kender (Group), a company devoted to auto-lubrication from Louth, Ireland (see resources), manufacturers of bearings prefer automatic lubrication. "Bearing manufacturers have long recognized the disadvantages of manual lubrication. The service life of rolling element bearings with automatic grease feed provisions ranks well ahead of most other means of lubricant delivery. Therefore, many process plants prefer automatic lubrication to traditional manual greasing."

Kender recommends several steps. Start off with a survey of your equipment and determine:

The number of lubrication points.
Type of lubricants used.
Optimum re-lubrication cycle.
Check suitability of either a single or multi-point auto lube system.
Installation requirements for proposed auto lube system.
Goals to be achieved.

Other procedures, worked out over years, will also have to be radically changed to take advantage of the new equipment. No one can be in all areas, so we will often have to enlist other groups such as operators, housekeepers, and security, to keep their eyes and ears open.

Predictive Maintenance

What do we mean by Predictive Maintenance? If we consult an authority, Merriam Webster Collegiate dictionary, we find the word predictive meaning "to declare or indicate in advance; especially: foretell on the basis of observation, experience, or scientific reason." It is from the Latin pre (before) and diction also from the Latin dicare "to proclaim." This definition fits closely with the maintenance concept of predictive.

The word 'maintenance' is somewhat more problematic since most of the definitions use the word, maintain. The third definition "the upkeep of property or equipment" works but does not shed much light. When we look up the word 'maintain' we hit pay dirt.

For the word 'maintain,' Merriam Webster Collegiate dictionary starts with "to keep in an existing state (as of repair, efficiency, or validity): preserve from failure or decline." One of the other definitions takes a different stab at maintain as in "to sustain against opposition or danger: uphold and defend"

Our goal in maintenance is to keep our physical assets in an existing state. Prediction is a declaration in advance that something is going to happen. From the dictionary 'Predictive Maintenance' is a proclamation or declaration in advance based on observation to preserve (something) from failure or sustain it against danger.

From this definition we can see at a very basic level that:

1) **Any inspection activity on the PM task list is predictive.** The reasoning is that at their most basic, all inspections look at an asset and decide if there is wear going on that will result in failure. The inspector then declares that such and such a bearing is squealing and is going to fail. That result is clearly predictive by this definition.

2) **Predictive Maintenance is a way to view data and does not necessarily require buying a lot of equipment.** The definition does not mention the means used. In fact it seems to be oriented toward reasoning and experience being applied to an observation. The important thing is that the conclusion be based on observation, judgment, and reasoning; and that determines if an act is predictive. In modern terms we would say this is a data issue. How we reason from the data determines the predictive nature.

By common agreement in our usage of the phrase 'predictive maintenance' we mean maintenance activity that includes some instrument or technology. Properly, any instrument can be used for predictive maintenance, if it is indeed used predictively (again predictive maintenance is about how you use the data). For example predictive inspections can come from existing equipment used in new ways including volt/ohm meters, meggers, measuring instruments, etc. All the predictive techniques should be listed on a task list and controlled by the PM system.

Why go through this dictionary exercise?

It's simple, I wanted to clear up the confusion about Predictive Maintenance. Predictive maintenance is not something you buy. **Predictive Maintenance is a state of mind!**

"Scientific application of proven predictive techniques increases equipment reliability and decreases the costs of unexpected failures." The meaning of this statement is very slippery. Many people take this statement literally and believe that predictive maintenance in itself extends life. Actually doing a predictive task such as an infrared scan doesn't extend life.

The scan may show a hot connection. Cleaning and retightening of the hot electrical connection done as a result of the scan does extend life. Predictive Maintenance is a maintenance activity geared to indicating where a piece of equipment is on the failure curve and predicting its useful life. Write-ups of the corrective items, the transfer of the deficiency to the backlog, and finally the completion of the work order, is what actually extend the life. Short repairs when the inspector is going on his or her rounds extend life.

The way predictive maintenance improves reliability is by detecting deterioration earlier than it could be detected by manual means. This earlier detection gives the maintenance people more time to intervene (hopefully enough time to intervene before failure!). Given the additional time, the corrective action can include ordering parts and materials (air freight not needed), planning, scheduling downtime, and straight time labor. With the longer lead time there is also less chance of an unscheduled event catching you unaware.

In the appendix to RCM II, John Moubray publisihed by Industrial Press which lists over 50 techniques for predictive maintenance. Every year some smart scientist, engineer, or maintenance professional comes up with one or two more. Any technique could be the one that will really help in detecting your modes of failure.

How is condition-based maintenance related to predictive maintenance?

In condition-based maintenance the equipment is inspected, and based on some specific condition, further work or inspections are done. For example, in a traditional PM program, a filter might be scheduled for change out monthly. In condition-based maintenance, the filter is changed when the condition of differential pressure (readings taken before and after the filter) exceeds a certain number of PSI. You might check a truck's oil level every time you fill up and top it off, but if the oil level is 2 quarts down you might initiate a series of low-oil inspections and other checks.

How is condition monitoring related to predictive maintenance?

Typically, in preventive or predictive maintenance we inspect an asset every day, week, month, or even less often. This procedure is effective because the duration from when the deterioration is detectable to when the critical wear causes a failure is longer then the inspection frequency. This critical concept is explained more fully in Chapter 18. There we discuss the P-F curve (performance verses failure curve), which is one of the basics of sophisticated preventive maintenance.

But what happens if this interval from detection to failure is short? What if it is one hour from good smooth operation through deterioration to failure? Traditionally items that failed that quickly were left off the PM program because there was nothing to be done. At the most a PM would be directed at eliminating dirt or getting at the cause of the failure in the first place (not a bad idea, in any event).

Predictive maintenance tools such as transducers of various types (temperature, acceleration) could be permanently mounted and monitored by, first PLCs, computer controllers, and eventually standard desktop computers. This full time monitoring could cycle at millisecond to microsecond speeds to pick up even the fastest deterioration. If certain readings were exceeded (such as temperature 10 degrees above ambient or acceleration over 2 Gs) either an alarm would sound or an automatic shutdown sequence would be initiated. The second advantage is that condition monitoring is non-interruptive, which means PM inspections are going on while the machine is making money!

Condition monitoring was once common only on very expensive assets like turbines. It is increasingly common as computers are added to almost all equipment for control. Adding condition monitoring just can be some new software, wiring, and a few transducers. As competition forces machine builders to offer more and more, these features become very attractive.

Borrowed

The ideal situation in maintenance is to be able to peer inside your components and replace them just before they fail. Technology has been improving significantly in this area. Tools are becoming available that can predict corrosion failure on a transformer; examine and videotape boiler tubes, or detect a bearing failure weeks before it happens. Many of these tools are borrowed from other industries.

Maintenance has borrowed tools from other fields such as medicine, chemistry, physics, auto racing, aerospace and others. These advanced techniques include all types of oil analysis, ferrography, chemical analysis, infrared temperature scanning, magna-flux, vibration analysis, motor testing, ultrasonic imaging, ultrasonic thickness gauging, shock pulse meters, and advanced visual inspection.

These technologies are invented or refined where the stakes are the highest. Some of the frequent sources are the military (oil analysis, infrared), medicine (ultrasound, miniature cameras), nuclear power plants (all kinds of non-destructive testing (NDT).

Look for a problem, which if solved, would make money or save lives. Consider the competitive advantage to a racecar team that has the ability (before any of their competitors) to determine if a camshaft has a crack (eddy current testing). Much of the R&D happens in medicine because the stakes are high and the monetary reward is also high. Look at ultrasound, chemical testing, NDT. Predictive Maintenance is all around in medicine.

Good place to outsource

Testing is a business of service bureaus, which come in all sizes. It is also one of the few areas where individual engineers, and skilled maintenance professionals can create a small business, make a good living, and provide services for all kinds of clients. Every technology has service companies that will provide anything from the equipment itself to training to use a turnkey service. Many have menus of services. Most metropolitan locations have service companies to perform these services, or rental companies willing demonstrate some techniques in your facility.

Baseline is basic

Almost all techniques depend on baselines to be most useful. This concept is similar to medicine. The doctor wants to see you when you are well. The doctor records readings of blood pressure, blood chemistry, and your physical exam. If you return later feeling sick the doctor can compare the readings. The simple act of comparison will simplify diagnosis of a disease from the presence of a normal variant.

The baseline readings are the readings when the asset is operating normally with no significant critical wear going on. In older facilities, getting the baseline is a significant problem (because everything is in some state of breakdown or loss of function). In many fields (such as air handlers, mobile equipment, generator sets, or motors) baseline data can be obtained from the manufacturer. In fact, the OEM for major items such as turbines require a full set of readings after installations. These measurements allow them to determine if the installation was done correctly.

Before you start a predictive maintenance program, consider these questions:

1. What is our objective for a predictive maintenance program? Do we want to reduce downtime, maintenance costs, or the stock level in storerooms? What is the most important objective? With any journey, knowing where you want to end up is useful!

2. Are we, as an organization, ready for predictive maintenance?
   - A. Do we have piles of data that we already don't have time to look at?
   - B. If one of the PM mechanics comes to us asking for a machine to be rebuilt, do we have time to rebuild a machine that is not yet broken?
   - C. Could we get downtime on a critical machine on the basis that it might break down?

D. Are we willing to invest significant time and money in training?
E. Do we have the patience to wait out a long learning curve?

3. Is (are) the specific technique(s), the right technique(s)?
   A. Does the return justify the extra expense?
   B. Do you have existing information systems to handle, store, and act on the reports?
   C. Is it easy and convenient to integrate the predictive activity and information flow with the rest of the PM system?
   D. Is there a less costly technique to get the same information?
   E. Will the technique minimize interference with our users?
   F. Exactly what critical wear are we trying to locate?

4. Is this the right vendor?
   A. Will they train you and your staff?
   B. Do they have an existing relationship with your organization?
   C. Is the equivalent equipment available elsewhere?
   D. For a service company, are they accurate?
   E. How do their prices compare with the value received, and to the marketplace?
   F. Can the vendor provide rental equipment (to try before you buy), can they provide a turnkey service giving you reports, and hot line service for urgent problems?

5. Is there any other way to handle this instead of purchase?
   A. Can you rent the equipment?
   B. Can you use an outside vendor for the service?

An excellent treatment of the whole field of predictive maintenance can be found in John Moubray's book *RCMII* published by Industrial press. In his work the technologies are grouped around detecting deterioration in the 6 effects:

Dynamic (vibration)	Physical (crack detection)
Particle (ferrography))	Temperature (infrared)
Chemical (water analysis)	Electrical (ampere monitoring)

In this work we group the technologies around chemical (including particle), mechanical (all dynamic modalities), and energy (temperature, electrical, optical and including miscellaneous).

The steel industry is not usually known for being early adopters of change. But the industry has had a wake up call. Tom McNeil, is Amex manager (All-Maintenance Excellence) at Gary Works. He described the change. "Preventive maintenance is when you change the oil in your car every 3,000 miles whether it needs it or not," McNeil says. "Predictive maintenance is when you sample the oil from time to time and check for any changes in its characteristics. You may find out you need to change the oil more often. It's a much more accurate maintenance technique and reduces costs by keeping you from discarding perfectly good equipment." Gary Works uses seven major diagnostic tools in its predictive-maintenance program on a regularly scheduled basis:

- Vibration analysis. "There are 1,800 machine trains throughout the steel plant," McNeil says. "They're checked monthly to detect any variations from the last reading."
- Thermography. More than 700 heat-generating points are checked each month, mostly with infrared equipment, to detect thermal anomalies.
- Fluid analysis. Each month employees take and analyze more than 800 samples of fluids from gearboxes, transformers, and other equipment.
- Visual inspection. Inspectors travel scheduled routes checking such things as the presence of coupling guards and the integrity of belts. Sometimes steel mills overlook this important maintenance tool, McNeil says.
- Operational-dynamics analysis. Using various devices, employees check equipment to make sure it's meeting design specifications. A damper might be checked to make sure it's receiving a 50-percent airflow, as designed.
- Electrical monitoring. Technicians regularly check all electrical components with voltmeters, infrared equipment, and other devices to guarantee their operational integrity.
- Failure analysis. These analyses determine why a piece of equipment failed and how that can be prevented in the future.

Chemical and Particle Analysis Predictive Tasks

One of the most popular families of techniques to predict current internal condition and impending failures is chemical analysis. There are 7 basic types of chemical analysis. The first two are related to particle size and composition:

Type	Material
1. Atomic Emission (AE) spectrometry	all materials
2. Atomic Absorption (AA) spectrometry	all materials
3. Gas chromatography	gases emitted by faults
4. Liquid chromatography	lubricant degradation
5. Infra-red spectroscopy	similar to AE
6. Fluorescence spectroscopy	assessment of oxidation products
7. Thin layer activation	uses radioactivity to measure wear

Oil analysis is a significant subset of all of the chemical analysis that is used for maintenance. The two spectrographic techniques are commonly used to look at the whole oil picture. They report all metals and all contamination, based on the fact that different materials give off different characteristic spectra when burned. The results are expressed in PPT or PPM (PPT-Parts per Thousand, PPM-Parts per million, PPB-Parts per billion).

The lab or oil vendor usually has baseline data for types of equipment that it analyzes frequently. The concept is to track trace materials over time and determine where they come from. At a particular level, experience will dictate an intervention is required. Oil analysis costs $10 to $25 per analysis. It is frequently included at no charge (or low charge) from your supplier of oil.

You are usually given a computer-printed report with a reading of all the materials in the oil and the 'normal' readings for those materials. Sometimes the lab might call the results in so that you can finish a unit, or capture a unit before more damage is done.

For example, if silicon is found in the oil then a breach has occurred between the outside and the lubricating systems (frequently silicon contamination comes from sand and dirt). Another example would be an increase from 4 PPT to 6 PPT for bronze, which probably indicates increasing normal bearing wear. This wear would be tracked and could be noted and checked on the regular inspections.

Oil analysis includes an analysis of the suspended or dissolved non-oil materials including Babbitt, Chromium, Copper, Iron, Lead, Tin, Aluminum, Cadmium, Molybdenum, Nickel, Silicon, Silver, and Titanium. In addition to these materials the analysis will show contamination from acids, dirt/sand, bacteria, fuel, water, plastic, and even leather.

The other aspect of oil analysis is a view of the oil itself. Questions answered include: has the oil broken down, what is the viscosity, are the additives for corrosion protection or cleaning still active? Consider oil analysis as a part of your normal PM cycle. Since oil analysis is relatively inexpensive, also consider doing it:

1. Following any overload or unusual stress
2. If sabotage is suspected
3. Just before purchasing a used unit
4. After a bulk delivery of lubricant to determine quality, specification, and if bacteria are present
5. Following a rebuild, to baseline the new equipment and for quality assurance
6. After service with severe weather such as flood, hurricane, or sandstorm.

Other tests are carried out on power transformer oil, and show the condition of the dielectric, and breakdown products.

In the words of Insight Services (see resources section) "The goal of an effective oil analysis program is to increase the reliability and availability of your machinery, while minimizing maintenance costs associated with oil change outs, labor, repairs, and downtime. Insight Services offers a lengthy battery of tests to assess the following three aspects of oil analysis: Lubricant Condition, Contaminants, and Machine Wear."

Lubricant Condition. The assessment of the lubricant condition reveals whether the system fluid is healthy and fit for further service, or is ready for a change.

Contamination. Ingressed contaminants from the surrounding environment in the form of dirt, water, and process contamination, are the leading cause of machine degradation and failure. Increased contamination alerts you to take action to save the oil and avoid unnecessary machine wear.

Machine Wear. An unhealthy machine generates wear particles at an exponential rate. The detection and analysis of these particles assist in making critical maintenance decisions. Machine failure due to worn out components can be avoided. Remember, healthy and clean oil leads to the minimization of machine wear.

The first place to begin looking at oil analysis is with your lubricant vendor. If your local distributor is not aware of any programs, contact any of the major oil companies. If you are a very large user of oils and are shopping for a yearly requirement you might ask for analysis as part of the service. Some vendors will give analysis services to their larger customers at little or no cost.

Labs that are not affiliated with oil companies exist in most major cities, especially cities that serve as manufacturing or transportation centers. Look for a vendor with a hot line service who will call Email or fax you if there is an imminent breakdown. These firms will prepare a printout of all of the attributes of your hydraulic, engine, cutting oils, or power transmission lubricants. The firm should be able to help you set sampling intervals and train your people in proper techniques of taking the samples.

Tip: Send samples taken at the same time on the same unit to several oil analysis labs. See who agrees, who's the fastest, which has the least cost. Pick a lab that maintains your data on a computer and be sure you can get the data for analysis.

Want to give oil analysis a try? Go to: *http://www.testoil.com/frame_freeoffer.html* they have a free offer to try out oil analysis.

Wear particle analysis, Ferrography and Chip detection

Particle techniques	**Description**
1. Ferrography	20-100 microns, ferrous only
2. Chip detection	40-microns up, metals only
3. X-Ray fluorescence	After radiation materials that emit characteristic x-rays
4. Blot testing	Blot highlights size and type of particles
5. Light detection and ranging	Analyze smoke from smoke stacks

These techniques examine the wear particles to see what properties they have. Many of the particles in oil are not wear particles. Wear particle analysis separates the wear particles out and trends them. When the trend shows abnormal wear then ferrography (microscopic examination of wear particles) is initiated.

Several factors contribute to the usefulness of these techniques. When wear surfaces rub against each other they generate particles. This rubbing creates normal particles that are small (under 10 microns), round (like grains of beach sand) and benign. Abnormal wear creates large particles that are irregularly shaped with sharp edges. All particles generated are divided by size into two groups: small <10 microns and large particles >10 microns.

When abnormal wear occurs, the large particle count dramatically increases. This is the first indication of abnormal wear. After abnormal wear is detected the particles are examined (ferrography) for metallurgy, type, and shape. These examinations contribute to the analysis of what is wearing and how much life is left.

The most obvious chip detection technology is a magnetic plug in the sump of an engine. You examine the plug to see if dangerous amounts of chips are in the oil. Chip detection is a pass-fail method of large particle analysis. Too many large particles set off an alarm. Several vendors market different types of detectors. One type allows the oil to flow past a low power electrical matrix of fine wires. A large particle will touch two wires and complete a circuit to set off the alarm.

Ferrography report: From Insight Services (a complete service facility for oil analysis. See resource section for contact information.

	10/17/95	9/18/95	8/17/95	7/21/95	6/20/95	5/19/95	4/18/95	3/17/95	2/21/95	
	10713	9733	8759	8173	7695	7283	6456	5786	5275	
SOOT	0.0	0.0	0.0	0.0	0.0	0.0	0.0	0.0	0.0	
SULPROD						0.1	0.1	0.1		
WATER EP	1.6		2.6	0.0	0.0			0.2		

PARTICLE COUNT

Date	10/17/96
LabNo	10715
> 2	27,987
> 5	11,872
> 15	2,138
> 25	987
> 50	54
> 100	9
ISO 4406	22/21/18

ANALYTICAL FERROGRAPHY

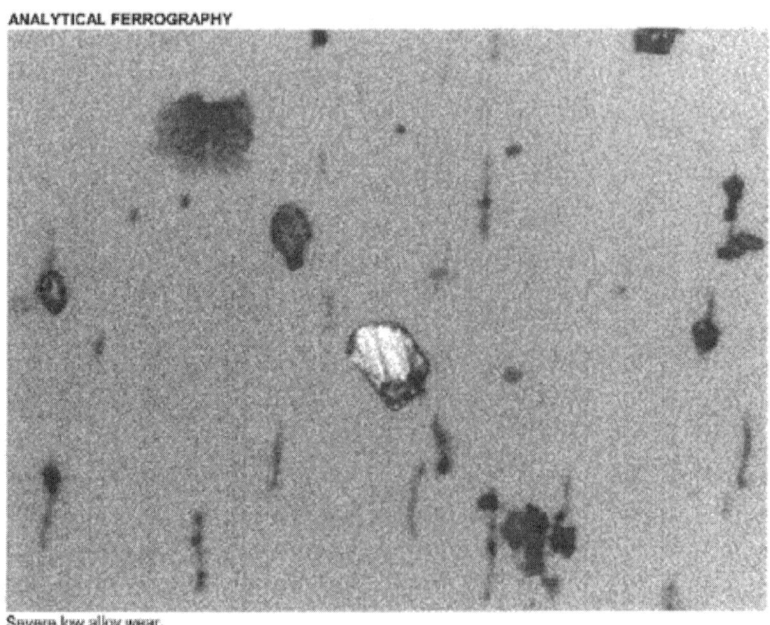

Severe low alloy wear.

Mechanical Predictive Tasks

Vibration Analysis

Vibration analysis is a widely used method in plant/machinery maintenance. A study in the city of Houston's wastewater treatment department showed $3.50 return on investment for every $1.00 spent on vibration monitoring. The same study showed that a private company might get as much as $5.00 return per dollar spent. The engineering firm of Turner, Collie, and Braden, of Houston, Texas, did the study and the vibration-monitoring project.

Each element of a rotating asset vibrates at characteristic frequencies. A bent shaft will always peak at twice the frequency of the rotation speed. A ball bearing, on the other hand, might vibrate at 20 times the frequency of rotation.

There are over 9 different types of vibration analysis. Each individual technique focuses on one aspect of the way assets deteriorate that is detectable by vibration. Techniques include octave band analysis, narrow band frequency analysis, real time analysis, proximity analysis, shock pulse monitoring, kurtosis, acoustic emission, and others.

The most popular is broadband analysis. This analysis measures the changes in amplitude of the vibration by frequency over time. This amplitude by frequency is plotted on an XY axis chart and is called a signature (measured for a given service load). Changes to the vibration signature of a unit mean that one of the rotating elements has changed characteristics. These elements include all rotating parts such as shafts, bearings, motors, and power transmission components. Anchors, resonating structures, and indirectly connected equipment, are also included.

Many large engines, turbines, and other large equipment have vibration transducers built-in. The vibration information is fed to the control system, which can shut down the unit or set off an alarm when vibration exceeds predetermined limits. The system also has computer outputs that allow transfer of the real-time data to the maintenance information system.

Effective vibration analysis requires a good deal of knowledge about the machine being measured and about the nature of vibration.

Quick set-up of a vibration-monitoring program:

1. Buy or rent a portable vibration meter.
2. Train mechanics in its use and make it a regular task on a task list assigned to the same person.

3. Record readings at frequent intervals. Transfer readings to a chart (or use a spreadsheet program and have it do the charting for you.
4. Take readings after installation of new equipment.
5. Compare periodic readings and review charts to help predict repairs.
6. Make repairs when indicated, do not defer. Note condition of all rotating elements; determine what caused any increase in vibration.
7. Review all vibration readings before and after you overhaul a unit, .
8. As you build a file of success stories, move into more sophisticated full spectrum analysis. Train more widely and trust your conclusions.

Frequency	Cause	Amplitude	Phase	Comments
1/2 rotational speed	Oil whip or oil whirl	Often very severe	Erratic	High speed machines where pressurized bearings are used
1 * RPM	Unbalance	Proportional im balance	Single	Check to loose mounting reference if in mark vertica ldirection
2 * RPM	Looseness	Erratic	Two marks	Usually high in vertical direction
2 * RPM	Bent shaft, misalignment	Large in axial direction	Two marks	Use dial indicator for positive diagnosis
1,2,3,4 * RPM	Bad drive belts	Erratic	1,2,3,4 unsteady	Use strobe light to freeze faulty belt for inspection
Synchronous or 2 * synchronous	Electrical	Usually low	Single or double rotating mark	If vibration drops out as soon as electricity is turned off it is electrical
Many * RPM	Bad bearing	Erratic	Many reference marks	If amplitude exceeds .25 mils suspect bearings
RPM * # gear teeth	Gear noise	Usually low	Many	
RPM * number of blades on fan or pump	Aerodynamic or hydraulic			Rare

Vibration causes - courtesy of August Kallemeyer's book Maintenance Management

One of the interesting differences between vendors is in the analysis. The hardware to collect and process the raw data improved very quickly (from the 1980s to the 2000s the capability increased and the price dropped by a factor of ten). Now the software is catching up.

For example WinProtect (Vibration Specialty Company) vibration data management can learn your vibration analysis routine and begin to think like you. It would learn whether a certain vibration characteristic signified a particular fault or condition. You can adjust the analysis engine to recognize and diagnose that condition. In their literature the company claims "With WinProtect, you can effectively "train" the analysis engine to better "understand" your machinery. As it gains understanding, it gains intelligence. It only gets smarter as time goes slowly by."

There is an excellent web site designed by Erik Concha (see resources section) to explain the field. He starts with the definitions of the basic terms: Acceleration - Velocity – Displacement – Phase, then goes on to more complex matters.

Tests, training, and certifications in vibration

As mentioned, vibration analysis requires a good understanding of both the engineering of the asset and the physics of vibration. As a result, training is available in most major cities.

In the US, for example there are tests, training, and certifications for knowledge of vibration analysis. Maintenance Technology Magazine (see resource section) has compiled a list of courses, tests and certifications. Some of these certifications have been actively available for almost 10 years.

Entry Vibration Analysis	Technical Associates of Charlotte (see resource section)
Analysis I, II, II Vibration	Technical Associates of Charlotte (see resource section)
Vibration Specialist I, II, III	Vibration Institute (see resource section)

Ultrasonic Inspection

One of the most exciting families of technologies is based on Ultrasonics. It is widely used in medicine and has also moved into factory inspection and maintenance. There are four or five techniques that make up this family.

In one of the most common and inexpensive techniques, an ultrasonic transducer transmits high frequency sound waves and picks up the echo (pulse-echo). Echoes are caused by changes in the density of the material tested. The echo is timed and the processor of the scanner converts the pulses to useful information such as density changes and distance.

Ultrasonics can determine the thickness of paint, metal, piping, corrosion and almost any homogenous material. New thickness gauges (using continuous transmission techniques) will show both a digital thickness and a time based scope trace. The trace will identify corrosion or erosion with a broken trace showing the full thickness and an irregular back wall. A multiple echo trace shows any internal pits, voids, and occlusions (which cause the multiple echoes).

Another excellent application of ultrasonic inspection is Bandag's casing analyzer, which is used in truck tire retreading. Ultrasonic inspection is used to detect invisible problems in the casings that could result in failures and blowouts after retreading. Ultrasonic waves bounce around from changes in density, so imperfections in casings (like holes, cord damage, cuts) are immediately obvious. The transmitter is located inside the casing and 16 ultrasonic pick- ups feed into a monitor. The monitor immediately alerts the operator to flaws in the casing.

A different application of Ultrasonics is in the area of ultrasonic detection. Many flows, leaks, bearing noises, air infiltration, and mechanical systems, give off ultrasonic sound waves. These waves are highly directional. Portable detectors worn like stereo headphones translate high frequency sound into sound we can hear so that we can quickly locate the source of the noises and increase the efficiency of the diagnosis.

Some organizations enhance this application with ultrasonic generators. The generator is inserted inside a closed system such as refrigeration piping or vacuum chamber and listen all around for the ultrasonic noise. The noise denotes a leak, loose fitting, or other escape route.

Energy Related Tasks and Miscellaneous Tasks

Temperature Measurement

Since the beginning of the industrial age, temperature sensing has been an important issue. Friction (or electrical resistance) creates heat. Temperature is the single greatest enemy for lubrication oils and for the power transmission components. Advanced technologies in detection, imaging, and chemistry allow us to use temperature as a diagnostic tool.

Temperature measurement is fundamentally different from other PdM technologies. The difference is in the clear and present danger that results from excessive heat. The failure of a motor control center is serious but not as serious as the ensuing fire. For this reason insurance companies have good data and sometimes require infrared scans of their electrical panels and distribution.

Hartford Steam Boiler Inspection and Insurance Co has proven through examination of their claims that a good infrared scanning program reduces fires and the associated loses. Their June 2001 article in Maintenance Technology demonstrates the effectiveness of the technology.

What is it?

Infrared photographic images are produced by heat rather than reflected light. Hotter parts show up as redder (or darker). Changes in heat will graphically display problem areas where wear is taking place or where there is excessive resistance in an electrical circuit. Infrared is unique since it is almost entirely non-interruptive. Most inspections can be safely completed from 10 or more feet away and out of danger.

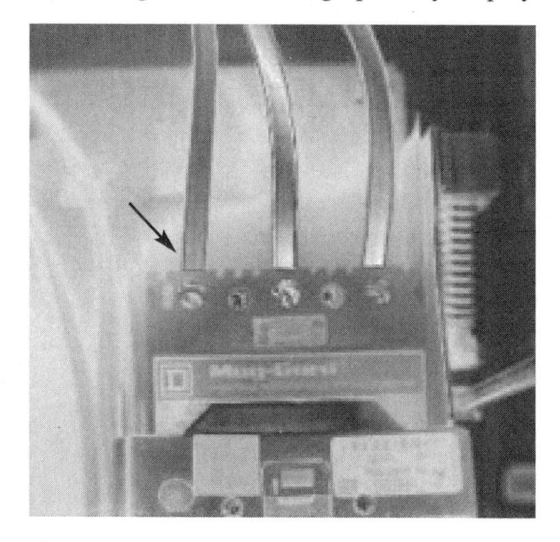

In the picture shown here (courtesy of Infrared Services) which is one of many great pictures on their web site and is reproduced with permission, one of three legs is hot (red-to-white) compared to the others. From this information we cannot tell what is wrong we just know something is wrong. Usually there is a loose connection, dirt and corrosion got underneath, and the resistance increased. In other examples, there could be a more serious fault in the machine, control circuits, or distribution system.

Readings are taken as part of the PM routine and tracked over time. Failure shows up as a change in temperature. Temperature detection can be achieved by infrared scanning (video technology), still film, pyrometers, thermocouple, fiber loop thermometry, other transducers, and heat sensitive tapes and chalks.

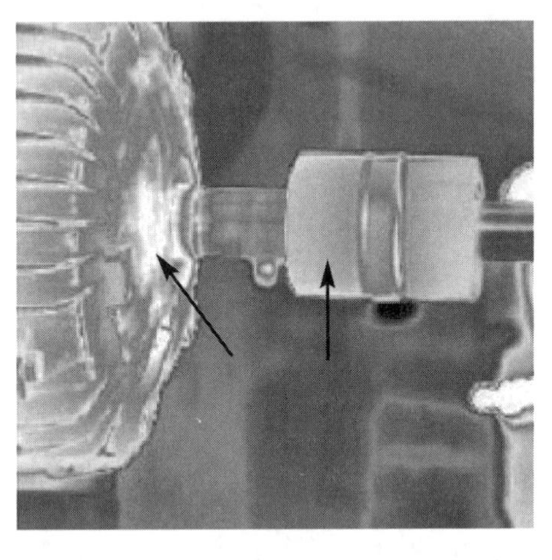

On large engines, air handlers, boilers, turbines, etc. temperature transducers are included for all major bearings. Some packages include shutdown circuits and alarms that sound if temperature gets above certain limits.

Another use is mechanical systems which heat up as they convert flexure to heat, convert friction to heat, vent process heat. With all systems, heat is transferred by convection (air currents), conduction (heat moves from a hotter area to a cooler area, and radiation. Infrared can be used to measure radiation. In the picture of the motor and bearing (Infrared Services) the bearing is misaligned and flexing. The flexing is causing the motor bearing to do extra work and heat up. You can also see the coupling on the motor side heating up, also caused by the misalignment.

The hardware for infrared is becoming more and more powerful. A typical specification for a top of the line, hand held imager with fully accurate, absolute temperature measurement capabilities might weigh 5 pounds, with 4X zoom, color viewfinder, self calibration, 30,000 hr MTBF, thermoelectric cooling, snap-in battery pack. Output (can go to a VCR, or to RS-232, and inputs include voice notations of each video image. Up to 500 images can be stored.

Harry Devlin, an Agema representative and infrared survey engineer, explains that the extra money gets you high accuracy, vital where an accurate temperature measurement is needed to detect the severity of the problem, and repeatability is needed so that you can count on the reading. He says that the high-end equipment is most appropriate for large facilities where there is a need to prioritize the findings (using the temperature gradient to identify immediate and high priority problems) or for industries where calibration and repeatability is an issue (such as nuclear power generation). An infrared gun that can take spot temperatures (without imaging capability) would set you back around $500-1500.

Thermography is a business of service bureaus. An example of a small service provider is Infrared Services in Colorado. Their chief thermographer has been in the field for 10 years working on tens of thousands of images. One problem of bringing any of the predictive high technologies in house is the expense of gaining the experience beyond the classroom learning. His firm offers projects (contact information in Resources section) such as:

- Roof Moisture Surveys, Low Slope Roof Evaluation
- Electrical System Surveys
- Building Envelope Performance Surveys (Heat Loss Detection)
- Wall Moisture Surveys
- Infrared Testing on High Voltage Equipment
- Mechanical & Equipment Surveys
- Steam / Utilities / Piping Surveys
- Infrared Inspections of Refractory
- Residential, commercial, and industrial energy audits
- Gyclo snow melt systems...

Should this be glycol? Yes

Possible uses for infrared inspection	To Look for
Bearings	overheating
Boilers	wall deterioration
Cutting tool	sharpness
Die casting/injection molding equipment	temperature distribution
Distribution panels	overheating
Dust atmospheres (coal, sawdust)	spontaneous combustion indications
Furnace tubes	heating patterns
Heat exchanger	proper operation
Kilns and furnaces	refractory breakdown
Motors	hot bearings
Paper processing	uneven drying
Piping	locating under ground leaks
Polluted waters	sources of dumping in rivers
Power transmission equipment	bad connections
Power factor capacitors	overheating
Presses	mechanical wear
Steam lines	clogs or leaks
Switchgear, breakers	loose or corroded connections
Three phase equipment	unbalanced load
Thermal sealing, welding, induction heating equip	even heating

Examples of areas where savings are possible from application of infrared:

A hot spot on a transformer was detected. Repair was scheduled off shift when the load was not needed, avoiding costly and disruptive downtime.

A percentage of new steam traps, which remove air or condensate from steam lines, will clog or fail in the first year. Non-functioning steam traps can be readily detected and corrected during inspection scans of the steam distribution system. Breakdowns in insulation and small pipe/joint leaks can also be detected during these inspections.

Hot bearings were isolated in a production line before deterioration had taken place. Replacement was not necessary. Repairs to relieve the condition were scheduled without downtime.

Roofs with water under the membrane retain heat after the sun goes down. A scan of a leaking roof will show the extent of the pool of water. Sometimes a small repair will secure the roof and extend its life.

Infrared is an excellent tool for energy conservation. Small leaks, breaches in insulation, defects in structure are apparent when a building is scanned. The best time to scan a building is during extremes of temperatures (greatest variance between the inside temperature and the outside temperature).

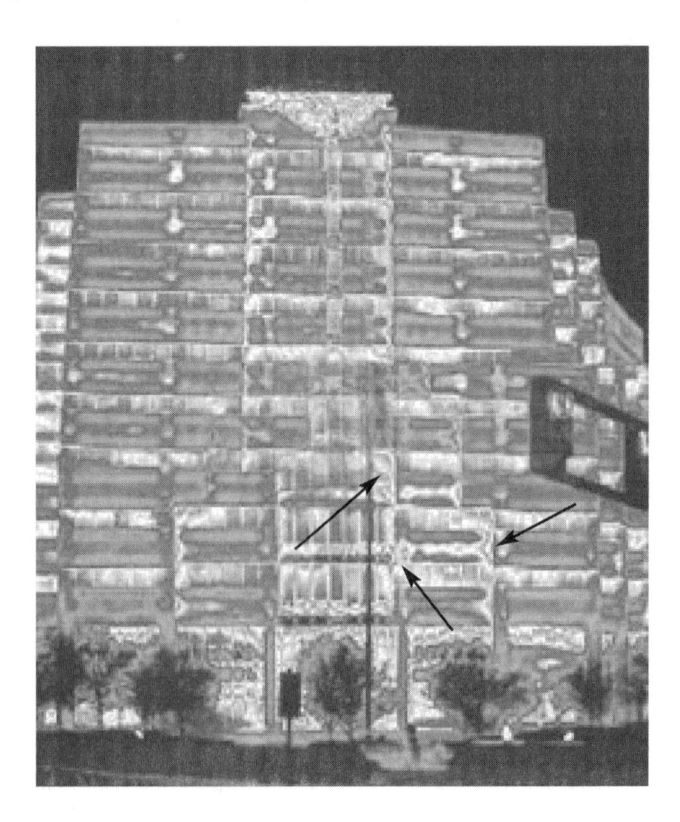

Furnaces are excellent places to apply infrared because of the cost involved in creating the heat and the cost of keeping it in place. Unnecessary heat losses from breaches in insulation can be easily detected by periodic scans. Instant pictures are available to detect changes to refractory that could be precursors to wall failures.

Temperature measurement in electrical closets is one area where experience can take the place of baselines. For example, in high voltage distribution, if the legs vary by 40 degrees C or more, then immediate action is required. If the legs vary by 10-15 degrees C then correcting actions can come at the next scheduled shutdown.

Tests, training, and certifications in infrared

Infrared is one of the most obvious technologies. A skilled electrician can get something from a scan of a panel without specific training. A little training goes a long way. Of course, as with all the predictive technologies there are many higher levels to the knowledge.

In the US, for example there are tests, training, and certifications for infrared knowledge, and skills. These courses include extensive hands-on use of the equipment. Maintenance Technology Magazine (see resource section) has compiled a list of courses, tests, and certifications. Some of these certifications have been actively available for almost 20 years.

Certified Infrared Thermographer Level I, II, III	Infraspection Institute

Level I, II, III Infrared Certification given by a variety of organizations includes 4-5 days of training: Snell Infrared, Infrared Training Center, Academy of Infrared Thermography,

Advanced Visual Techniques

The first applications of advanced visual technology used fiber optics in borescopes. In fiber optics, fibers of highly pure glass are bundled together. The smallest fiber optic instruments have diameters of 0.9mm (0.035"). Some of the instruments can swivel to allow the instrument to see the walls of a boiler tube. The focus on some of the advanced models is 1/3" to infinity. The limitation of fiber optics is length. The longest is about 12' though some are longer. The advantages are low cost (about 50% or less of equivalent video technology), and level of technology (they don't require large amounts of training to support them).

Another visual technology gaining acceptance is ultra-small digital video cameras. These cameras

are used for inspection of the interiors of large equipment, boiler tubes, and pipelines. These CCD (Charge Coupled Display) devices can be attached to a color monitor through cables (some versions used on pipelines can go to 1000'). A miniature television camera smaller than a pencil (about 1/4" in diameter and 1" long), is used, with a built-in light source. Some models allow small tools to be manipulated at the end; others can snake around obstacles. The equipment is extensively used to inspect pipes and boiler tubes.

The model shown on the prior by Inuktun Services Ltd (see resources section) can navigate 6" pipe and is able to penetrate up to 450 meters (1500 feet) of pipe, and overcome obstacles and offset joints. They have other motorized models for pipe as small as 4". In the parallel configuration (pipe larger than 300 mm / 12 inch), the vehicle is steerable, allowing it to easily maneuver through bends and joints. With the addition of a scissors hoist, the camera can be remotely raised to a height of 105 mm / 42 inches.

The major disadvantages are cost and level of support (the equipment requires training to adjust and use). The major advantage lies in flexibility. The heads or cables can be replaced and you can end up with several scopes for the price of one. If you have a good deal of piping this type of equipment is essential.

In most major industrial centers, service companies have been established to do your inspections for a fee. These firms use the latest technology and have highly skilled inspectors. Some of these firms also sell hardware with training. One good method is to try some service companies and settle on one to do inspections, help you choose equipment, and do training. You can also rent most of the equipment.

Other related visual equipment includes rigid bore-scopes, cold light rigid probes (up to 1.5 m), deep, probe endoscopes (up to 20 m), and pan-view fibrescopes.

Other Methods of Predictive Maintenance

Magnetic Particle Techniques (called Eddy Current Testing or Magna-flux)

Magna-flux is borrowed from automobile racing and racing engine rebuilding and has begun to be used in industry. This technique induces very high currents in a steel part (frequently used in the automotive field on crank and cam shafts). While the current is being applied the part is washed by fine, dark- colored magnetic particles (there are both dry and wet systems). Below is a suggested PM inspection from National Industrial Supply (see resources section) for an inspection of hooks.

This inspection pertains to magna-fluxing **crane hooks, hoist hooks, jib hooks, and lifting assemblies** for cracks. The inspection consists of the following:

1. Each hook or assembly (area to be inspected) will be cleaned and visually inspected for any apparent deformation or excessive ware.

2. Each hook or assembly (area to be inspected) will be measured for current dimensions, such as, but not limited to, the hook's throat opening, the hook's saddle thickness, and or the lifting pin's diameter, etc.

3. Each hook will be tested for cracks with a magna-flux or die-penetrant test.

4. A typed report listing a description of all the hooks and lifting assemblies inspected will be provided to the customer.

The test shows cracks that are ordinarily too small to be seen by the naked eye and cracks that end below the surface of the material. Magnetic fields change around cracks and the particles outline the areas. The test was originally used when re-building racing engines (to avoid putting a cracked crankshaft back into the engine). The high cost of parts and failure can frequently justify the test. The OEM's who built the cranks and cams also use the test as part of their quality assurance process.

Penetrating Dye Testing

Penetrating dye testing is visually similar to Magna-flux. The dye gets drawn into cracks in welded, machined, or fabricated parts. The process was developed for inspection of welds. The penetrating dye is drawn into cracks by capillary action. Only cracks that come to the surface are highlighted by this method.

Management of PM Activity

PM requires precise management to get the most bang for your buck. Since PM jobs tend to be shorter in duration it would be easy to have a work to non-work ratio of 1:10 or greater, which translates to 6 minutes of PM work per hour. The rest of the time is being spent traveling, getting permissions, locating materials, etc. If any re-source is missing, the PM cannot proceed. To insure no resource is missing the PM work must be completely planned.

Planning for PM

PM should be planned because then the return for your time invested is excellent due to the repetitive nature of PM. "The preferred approach is to prepare a thorough Planned Job Package for any repetitive job encountered. By this means the planner work-load is reduced as reference libraries are established". From *Maintenance Planning, **Scheduling and Coordination*** by Nyman and Levitt.

What is in the planned job package for a PM?

- Work order or PM ticket. Ticket should have blank space for write-up of short repairs accomplished during the PM
- Detailed Task list with step-by-step procedures for each task. This is especially important with PMs because they can be recurring frequently. Any PM should be totally thought through
- Estimated Labor hours by craft and skill.Consider contract as well as in-house resources.
- Bill of Material. List all materials needed for the job
- Requisitions for materials not in stock
- Complete list of tools required. In the PM review process, determine if different, specialized, custom, better, or electric tools will speed the job.
- A complete list of the required permits, clearances and tag outs is included in the planning package. Note that the responsible mechanic or equipment operator must take the final steps at the time of scheduling.
- List of safety requirements including lock-out, confined space, fall protection and personal protective equipment

Equipment access requirements. List who has to be notified when maintenance takes custody of the asset. If necessary, where are the keys?

Service manuals, prints, sketches, digital photos, or Polaroid's, special procedures, specifications, sizes, tolerances, and other references that the assigned crew is likely to need

Blank write-up forms for corrective jobs to be entered back into planning.

Scheduling PM

Listing. From the active planning effort you have a complete listing of everything needed for the PM to take place. That list is step one of a three-step process.

Scheduling is " bringing together in precise timing the six elements of a successful maintenance job: labor, tools, materials, parts, supplies, information, engineering data and drawings, custody of the unit being serviced, and the authorizations, permits, and statutory permissions". Also from the book **Maintenance Planning, Scheduling and Coordination** by Nyman and Levitt. This precise merging of all resources required for a job is particularly essential in PM activity.

Coordination. Within the scheduling umbrella is coordination. Coordination with production is the second step. This step tries to mesh your wishes of when to perform PM tasks with their production realities.

To effectively coordinate you must create a schedule proposal. This proposal is presented to the production or operations group to see if the time slots when you have personnel and materials available coincide with the slots that can provide the equipment. After the wrangling is done you can proceed to the third step which is taking your proposed schedule (the one that production has agreed to) and formalizing it.

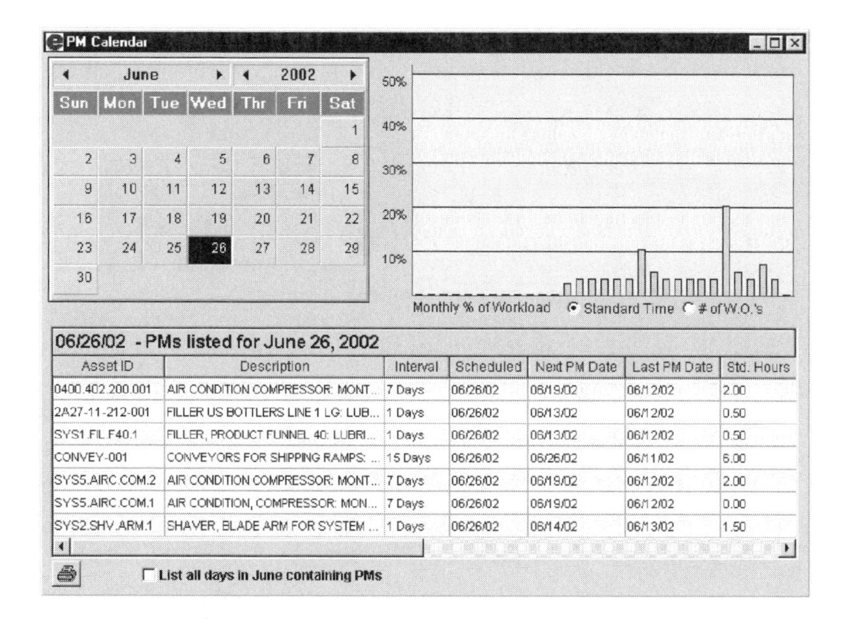

Asset ID	Description	Interval	Scheduled	Next PM Date	Last PM Date	Std. Hours
0400.402.200.001	AIR CONDITION COMPRESSOR: MONT...	7 Days	06/26/02	06/19/02	06/12/02	2.00
2A27-11-212-001	FILLER US BOTTLERS LINE 1 LG: LUB...	1 Days	06/26/02	06/13/02	06/12/02	0.50
SYS1.FIL F40.1	FILLER, PRODUCT FUNNEL 40: LUBRI...	1 Days	06/26/02	06/13/02	06/12/02	0.50
CONVEY-001	CONVEYORS FOR SHIPPING RAMPS: ...	15 Days	06/26/02	06/26/02	06/11/02	6.00
SYS5.AIRC.COM.2	AIR CONDITION COMPRESSOR: MONT...	7 Days	06/26/02	06/19/02	06/12/02	2.00
SYS5.AIRC.COM.1	AIR CONDITION, COMPRESSOR: MON...	7 Days	06/26/02	06/19/02	06/12/02	0.00
SYS2.SHV.ARM.1	SHAVER, BLADE ARM FOR SYSTEM ...	1 Days	06/26/02	06/14/02	06/13/02	1.50

Scheduling itself is divided into three activities:

- *Job loading,* listing the jobs to be done consistent with hours available
- *Job scheduling,* to spread the list of jobs out over the week taking into account estimated hours and skills available each day.
- *Manpower commitment,* to assign specific workers to individual jobs.

The CMMS can make PM management and scheduling easier. The following screen from eMaint (contact information in resources section) dramatically improves the ability of the planner (or anyone in that role). In the example shown, the computer can look ahead to any date and display all PMs due on that date (it can even predict when meter based PMs might become due. It can calculate all the standard times for all PMs that are due and show the percentage to available hours. The system also lists the specific PMs available.

This is all well and good. What if production won't give us the equipment?

In spite of all of the articles and all the pleading, some equipment is unavailable for PM. This issue is called access to equipment.

One of the most difficult issues of maintenance is access to equipment (because the customer wants it or needs it). Access problems fall into two categories, political and engineering.

Political access problems are problems that stem from political reality. The equipment is not in use 24 hours, 7 days. But it is in use whenever you want it for PM. The reason you are not given access might have to do with production control having assigned no time for the PM, the maintenance department might be distrusted by production, etc. Some ideas for political access problems:

1. Go back to the planning department to discuss requirements. Do not wait until you need the unit the next day. Sometimes the production schedule is set weeks or months ahead of time. Lay out your PM requirements for each asset for a year in advance including the hours of down time. Dr. Mark Goldstein says that the 52 week PM schedule is essential and should be loaded into the Production Control system (MRP-Materials Resources Planning) as a constraint. In other words, PM is a ground rule just like the 15 minutes needed to change color in another system.

2. Your PM requirements are distrusted. You have to haul in respected outsiders, or circulate PM success stories from your plant or from the trade press to everyone in production management. Or it may be necessary to hire a respected consultant to take a look at your operation and render an opinion. Sometimes an outsider will say the same thing you've said for years and be listened to! Keep doing it until they believe you.

3. Always collect broken parts and build a display. Conduct a class in PM and

breakdown with these broken part collections and show how PM could have avoided the problems.

4. Use production and downtime reports, now in circulation, and highlight downtime incidents that could have been avoided by PM effort.

5. Most importantly, conduct your business with production with integrity. When you do get a window for PM or corrective work, give equipment back when promised, show up when promised, if there is a complication communicate with everyone; before, during, and after.

When it absolutely positively cannot be shut down

We call these engineering access problems. Engineering access problems are easy to spot. These access issues stem from equipment that cannot be taken out of service because it is always in use. Consider transformers, continuous process components, environmental exhaust fans, and single items without redundancy in this category.

Think about the main substation in your facility or the transfer switch feeding a hospital's ICU. What about a single compressor in a 24/7 manufacturer? These items cannot be shut down without extreme planning and scheduling. In general you must shut the whole plant or area down to service this equipment, or order or rent expensive back-up systems.

A partial antidote for both political and engineering access problems might be in non-interruptive maintenance.

Non-interruptive maintenance is defined as tasks that can be safely done without interrupting the normal operation of the machine. An example would be infrared scanning of a motor control center.

Interruptive/Non interruptive

This is a variation on the unit-based theme for machines that run 24 hours a day (or are running whenever you need to PM them). The unit-based list is divided into tasks that can be done safely without interrupting the equipment (readings, vibration analysis, adding oil, etc.) and tasks that require interruption. The tasks can be done at different times. The interruptive list may require half as much downtime as the original task list. It is also interesting to note that many of the high tech inspections require the asset to be running. The technology is all moving toward non-interruptive maintenance modalities.

Interruptive tasks	Non-interruptive tasks
Check coupling tightness	Scan all vibration points
Verify tightness of electrical connections	Check anchor bolt tightness (sometimes)
Jog machine to see alignment	Infrared scan of connections
Clean out jaws area	Sweep up around machine (sometimes)

Some tasks neatly divide into the two categories. If you can't have control of the asset the non-interruptive list might a good approach to PM. You end up with two task lists for each asset in this category.

Advantages: Reduced machine downtime
Disadvantages: Slightly less productive since the machine requires at least two trips

The next logical step- re-engineer the tasks

Sometimes the task can be re-engineered so that it can be done safely without machine interruption. The easiest example is lubrication.

Case Study

We completely cleaned the Centerless grinder for the first time in years. As part of this complete cleaning we disassembled the machine and examined it in comparison with the assembly drawing. The machine had 15 lubrication points, of which 6 had never been seen before. There were points up underneath the bed, and behind the motor. It seemed to have points all over the place. The shortened service life could probably be explained by not lubricating those points.

About 8 hours of piping solved this problem. All the points were piped to a central point where the operator could add grease without impacting production. An automatic lubrication system could have been installed because most of the work was already done.

Some systems like the Centerless Grinder lend themselves to re-engineering.

Another approach is the substitution of an interruptive task with a non-interruptive one. This substitution may require adoption of technological solutions. In the task list above consider the two tasks:

Interruptive task	Non-interruptive task
A. Check coupling tightness	B. Scan all vibration points

It might be possible to substitute task B for task A and eliminate the interruption. We have to determine if the two tasks are equivalent. We also have to determine if there is sufficient depth on the bench to carry out the more sophisticated or technological task. If we had an existing vibration program there would be no problem. The re-engineering would have substituted one task for another.

The third technique of re-engineering goes a step further in that it uses technology to not only inspect the equipment but also to decide if the equipment is healthy. This is an increasing trend of top end, very expensive, equipment.

Case Study

Cigarettes are manufactured on a high-speed assembly line. The line makes the cigarette, makes the pack, and fills it with cigarettes, makes the carton and fills it with packs, and makes a case and fills it with cartons. At the time of this case study the line was producing about 4,000,000 cigarettes or almost 1700 cases per hour. There were 5 lines in the building of which one was down for rebuild.

There was limited production capacity so that every cigarette made had a market. You can imagine that with the legal climate, few firms are investing in extra capacity (unless they can make their money back quickly). The line was run 24/7 - 365 except when it was down (both scheduled and unscheduled) or when they did the complete shutdown and refurbishment. That amounted to 8760 hours per year, less 300-400 hours a year from various kinds of downtime (about 96% uptime).

Rough downtime calculations were as follows:

$0.80/pack to the company or $0.04 per cigarette less materials and factory overhead of $0.02. In one hour the line throws off 4,000,000 * $0.02 or $80,000 gross profit before taxes

The line had a PM that took 5 hours per month. The PM time only resulted in losses of over 8300 cases of product or $400,000 every month. Needless to say, management was eager to get that PM time back. They were afraid of increasing downtime if they just cut out the PM.

The solution?

When the line was scheduled to be out of service for a complete (routine) rebuild the maintenance department jumped in and executed a well-wrought plan for PM task re-engineering. Some of the things they did were to add sensors to every critical bearing for temperature and acceleration; they even mounted a high-speed camera inside the line so that they could observe what was happening in real time. Automatic lubrication was added or upgraded to the state of the art.

All this technology was wired to a monitoring console that took readings every 50 milliseconds. After everything was done the PM went from 5 hours to 38 minutes. The 4-hour plus savings paid for the investment in engineering and work in 4 months (about $1,600,000).

Using Metrics to Manage PM and PdM

Depending on which magazine you read and when the article is written you'll be likely to see discussions of benchmarks and KPIs (Key Performance Indicators). The goal of both is the same, to quantify the performance of a unit so it can be measured, displayed and managed.

Benchmarking talks about historical benchmarks, best in class, and best in the world. Each of these is designed to give useful information to manage your area, department or business unit.

KPIs have a different basic aim in that they are designed to identify the key indicator that would measure what is going on in an area. Unlike benchmarks they are not comparisons at their core but measurements in themselves.

Ken Bannister, in an April 2002 article in PEM magazine distinguishes between KPIs for strategic purposes (organizational measures like profit), operational KPI for a business unit (such as plant downtime) and personal KPIs (such as individual machine uptime). All the measures below are operational or personal.

PM Compliance

This is a basic metric and should come out with every new PM schedule for the prior period. Transfer of this number every month to a trend chart can be used to detect decay in the PM priority.

Number of PM or PdM Tasks completed / Number of PM or PdM tasks scheduled

Raw PM Measurement

Sometimes simpler is better because everyone can understand it. This most obvious metric is the basic answer to the question of how much PM are we doing. Raw PM simply measures how much PM is done compared with overall hours.

Preventive maintenance hours (PM): % PM

Planned verses Unplanned

Planned work and PM systems are closely related but not identical. An effective PM system will promote a planned environment but may not be sufficient to create a Planned environment. If you want to look at the whole related picture of unplannable verses plannable take a look at:

Preventive maintenance hours (PM)	% PM	plannable
Emergency hours	% Emergency hours	un-plannable
DIN (Breakdown-Do it now) hours	% DIN	un-plannable
Short repair hours	% Short repairs	plannable
CM (Corrective maintenance) hours	% CM	plannable

The important breakdown is planned (PM+CM+Short repair+Project) to unplanned (DIN+EM). This ratio shows how much your facility is dominated by unscheduled events. The trends of these numbers give you a feel for whether there is improvement.

To look at this subject with more rigor we would dissect the planning effort and look at amount planned, planning compliance, effectiveness, and efficiency.

PM Effectiveness (PM work order hours/PM standard hours)

Where there are good standards on most PM/PdM jobs the effectiveness ratio can be useful. It shows how much work is really done. You get 3 hours credit for 3 standard hours even if the job took 16 hours.

Total backlog by craft (in hours, weeks per person)

Total and available backlog numbers are essential to run an effective maintenance effort. When a PM program is installed, the total backlog should shoot up. As the crews deal with deferred items the backlog should drop week to week and stabilize at a lower level. Working through the backlog is a good job for contractors because the workload after the backlog is worked off is lower.

Breakdown report

To improve, we consider that every breakdown is a failure of the PM program. It is important to review the breakdown report frequently, but it is even more important to use it in depth to improve the PM task lists.

Breakdown reports can take many forms. A list of breakdowns with causes is a minimum. Adding response times, and calculating MTBF (mean time between failures) with MTTR (mean time to repair) information, is useful for the PM failure mode analysis. A breakdown should always be treated as an educational opportunity to see where (if at all) the system failed.

Outsourcing PM

Outsource: Hire an outside firm to do the PM activity and the corrective maintenance that results from the inspection.

One common way to have the advantages of a PM program without the disadvantages is to outsource all PM, or outsource PM related to a certain asset group or class. With outsourcing the vendor does the PM checks, creates the write-ups and does the corrective maintenance.

The most common applications of outsourced PM includes:

Buildings, venues and facilities	HVAC, elevators, escalators, fire safety	Lack of in-house skill, liability
Hospitals	Biomedical, X-ray, MR equipment and above	Lack of skill, liability
Factories	Water treatment	Specialized knowledge and lower cost
Everywhere	Large computer systems	Mostly lack of skill and a rapidly changing environment

There are many reasons to outsource maintenance work and PM in particular. In most instances of successful outsourcing there are multiple reasons (such as saving money and improving quality).

The primary reasons to outsource PM or corrective maintenance would include:

- Save money, the contractor can do PM work less expensively than you.
- Improve quality by using specialists.
- Lack of skill within your crew (alarms, hi-tech maintenance, etc.).
- You don't have enough work to justify hiring that skill. You don't have enough equipment to justify keeping someone up on new techniques, skills, and ideas.
- Lack of specialized knowledge (related to above), including things like predictive maintenance and interpreting reports from more exotic inspections
- Lack of appropriate license (even if you have the skills)
- Lack of specialized and expensive equipment and tools
- Reduced legal liability (elevators, escalators, fire systems).
- Reduced hazard to own employees (PM inspection in tanks).
- You want an outside opinion (for political reasons) or you need an outside 'expert' to show you a whole new approach (what you've been doing has not worked at all).
- Training (send your mechanic(s) to help, or tag along to improve skills).
- Save time when you are already busy on other work. Or you are afraid the PM work would get shunted aside and not done.
- Don't want to manage the PM job (hiring the contractor to do that).
- You are not sure your systems are reliable enough so you rely on someone whose pay check depends on them showing up for the PM.

A description of a PM program for electric motors by Precision Electric In. shows clearly the advantages of using a specialist for PM work. The subject is common failure modes of motors:

"The most common cause of failure in an alternating current motor is mechanical."

"For this reason our primary concern for these units is mechanical condition. During operation, alternating current motors are tested using vibration analysis. High frequency G's are recorded as indicators of bearing condition. Additional mechanical testing includes displacement (mils) data, and velocity (in/seconds) data. The combination of all readings provides an excellent overview of the operating mechanical condition of the unit under test."

The next step discussed is creating a baseline and keeping trend data to uncover subtle deterioration:

"By performing these tests on all motors, a baseline can be set up and utilized from which to determine priorities of when units should be removed for routine service."

An in-depth analysis that would be unusual in all but the most sophisticated factories follows:

"In addition to vibration analysis data, the D.C. motors are visually inspected for cleanliness, commutation, brush wear, brush spring condition, and brush holder condition. The windings are tested dialectically using a 1000 VDC megohm meter. The information is recorded for future reference. Carbon brushes are identified and replacement part numbers are recorded. Because these procedures require hands-on contact with normally energized portions of the motors, they can only be performed during a machine shutdown. The shutdown is very brief and the work is scheduled to suit production demands…."

Finally the company's in-depth motor engineering ability and deep motor corrective maintenance capabilities are covered:

Precision Electric will also make recommendations in areas such as periodic blowing out of the units, brush replacement, bearing lubrication, cleaning of filters, and possible application problems. The purpose of this program is to reduce or eliminate unexpected, major, and expensive downtime. In the initial months an investment will be made into electric motor repair that will consist of clean up, bearing changes, turn and undercutting of commutators, changing leads, varnish treatment of windings, replacing brushes and complete dielectric and mechanical testing. What you will save will be the costly repairs that include all the above mentioned plus armature rewinds, commutator and brush holder replacement, field rewinds, stator rewinds, major machine work, and damage so extensive that the cost of repair may not be economically feasible. In addition to costly repair, you will be reducing or eliminating the even more costly by-product of motor failure, unexpected downtime." Adapted from Precision Electric, Inc. Mishawaka, IN.

The literature highlights the reasons that organizations outsource any work.

As with all outsourcing there are reasons to stay away from outsourcing PM…

1. In using outside contractors it is essential to define the scope of the work. The scope of work is easy in PM because the PM task list is supplied by the contractor or by yourself. The tough call in PM is that you have no easy way of knowing if any PM was done. How will you know if you are getting your money's worth?

2. Sometimes there is a negative public relations issue. Is there a negative image to using contractors when they are representing you to your customer/tenants/users?

3. There are as many ways to get ripped off, as there are people. Some possibilities include just not doing the PM work, possible quality problems where the contractor knowingly cuts corners, mistakenly hiring a con artist, or a well-meaning contractor who is out of his/her area of expertise in your job.

4. Dependency. You can become dependent and can't make a move without the contractor.

...and pitfalls to avoid.

1. Be careful of loose specifications (like just "PM Machine," what does that mean?).

2. In PM don't always take the lowest bid.

3. Take some time to define a "good job." Without a definition of performance, no one knows or can define what a good job would look like. Do accept clauses like "all work is expected to be done in a professional and workmanship manner, or All work will be in compliance with applicable city building codes."

4. Take some time to define the skill sets the workers should have. Even a good task list in the hands of a novice is almost useless (and it is not what you are paying for).

5. Particularly in public spaces and public buildings there must be statements about how the site is to be left at the end of each work day and how employees and the public are to be protected from the PM job. If they are on-site in off-hours, who is responsible for locking up, cleaning, and debris removal?

6. There should be deduction clauses that spell out what you will charge back and when you will charge it. Examples would be debris removal, clean-up, missing PM dates, etc.

7. There should be a clear cancellation clause. You need to spell out how and why you can cancel the contract.

8. You must have a schedule of extras. A common ploy is to low-ball the bid to get the PM job, then flood the organization with high charges for even the smallest short repair and corrective items. Look for clauses like " all extras not included in the original price have to be agreed to in writing prior to the commencement of the work." Another strategy is to develop a list of short repairs and corrective items with prices.

9. To be fair to both sides pick a contract term that is not too short or long for your needs.

Task List and Analysis

In the analysis of task lists there are two approaches and a third item that must be taken into account to insure that all areas are covered and no resources are wasted.

1. We want to be ruthless about eliminating tasks that do not promote reliability or are not 'worth it.'

2. We also want to be able to add tasks where the failure rate without them is unacceptable or there is a risk we are unwilling to take.

3. The third item is that inspection tasks must be chosen with the appropriate sensitivity. We want to make sure the task does not give us false positives (showing a problem when there isn't one) or false negatives (not showing a problem that is present).

In the section on RCM we discussed E (environmental) and S (safety) consequences. In the following discussion about task list analysis these E and S consequences take on a special significance. The goal of RCM analysis is to reduce the probability of E or S consequences to zero.

We inspect the hook from an overhead crane for distortion at the beginning of every shift. If the hook were used correctly we would never expect to find a damaged hook, but the inspection must be part of the shift startup procedure. The reason is that the hook inspection has an S consequence. With E or S consequences we want to reduce the probability of that type of failure to zero (as close as possible). If you are facing tasks that are designed to mitigate E or S consequences, be very careful of purely economic or operational analysis.

Four dimensions

Another way to look to look at this subject is to go back to the beginning of the book. Effective PM has four dimensions. The dimensions lead us to ask questions about the task. Each dimension contributes valuable information in the analysis of the task. At some point the different data provided by the analysis have to be weighed and the task accepted, rejected, or modified.

Every PM task has costs associated with it, including include labor costs, material costs and other costs (such as special tools, special PPE, overhead). So any item on a task list rolls up costs every time the PM is executed (once a week, month, or other interval). This is an essential aspect of PM – that costs are rolled up every week for years. A poorly chosen PM task will saddle your maintenance efforts with unnecessary additional costs for years or decades.

TASK on Sump Pump: Frequency – Quarterly: (Visually) Check bail, floats, rods, and switches. (Implied task-Touch all parts to make sure they are still up to the job) Fill sump with water and make sure float operates as designed. (Implied task - observe complete operation of sump pump) Clean sump and areaPossibility of short repair if something is loose bent or damaged. 30 minutes by mechanic US$20.00 No parts Water hose and water	**Engineering:** Makes a complete check. P-F curve shows corrosion or fatigue failure is mostly slow enough to be caught by a quarterly inspection. Mechanism will not operate if large debris falls into sump and jams float. Consider reengineering with a grate over the sump area if one is not already present. **Management:** PM can generate this task. Since it is in a remote area, task should be occasionally audited to insure it has been done. Good opportunity to equip the inspector for a short repair. **Economic:** This pump is protecting whole lower level from water. What is that worth? What is the probability of flooding? Has it ever flooded? If the lower level has expensive equipment or materials consider a back-up sump pump (or two) on separate electrical services. This pump is a good example of a task that will extend the life of the asset but that benefit in itself does not justify the $25-$50 per year. **Psychological:** Task can be in a dirty area, remote from everyone else. On the other hand it is an active task protecting the inspector from maximal boredom.

What does "worth it" mean?

Lets say we have a drill press in the maintenance shop that is used occasionally. We also have drill presses in the machine shop. A PM task to completely clean and lubricate the press will clearly extend its life. If this task takes 45 minutes a week is it worth it?

At US$40/hour ($40 represents fully burdened labor costs) the task is worth US$30 a week or $1560 per year. Only you can evaluate if $1560 is "worth it." One maintenance manager had a lottery every December and gave the drill press away to a member of the crew! What he found was that the maintenance workers were more careful with a drill press that they might get to keep.

Hotels have done the same thing with their housekeeping staff and vacuum cleaners. In that example, instead of PM on the vacuum cleaners the housekeeper could take the vacuum home if it lasted a whole year. In the past the vacuums had only lasted 4-6 months. The vacuum cleaner costs dropped dramatically.

The ball game is entirely different if the drill press is part of a closely linked manufacturing cell. Then the drill press performs an essential part of the process and without the drill the whole cell goes down. Losses could be hundreds or thousands of dollars an hour. Where the overall costs are high, the task is "worth it."

A good rule of thumb is that the cost of the PM should be less than 70% of the costs avoided in that year. In other words, if the PM cost is $1000 in that year then the avoided breakdown costs should be in excess of $1430 in the same year. If an asset had broken only every other year then we can only use 50% of the cost to compare with the PM cost.

Workshop for existing task lists

Team	Assemble a team that includes at least some "old-timers" who work with the equipment, engineers' familiar with the equipment, supervisors, parts buyers, and operators. The longer and better their memory the more effective is this analysis.
Step 1	Make a list of every failure any of you have seen. Use brainstorming techniques. Don't analyze at this stage, but go around and make as complete a list as possible. List each failure on a separate sheet (or better yet, on a computer file in a word processor or spread sheet).
Step 2	Track back the failure as far as possible to its cause. We can readily see the result of a smashed impeller or fried motor starter. Can we look back at the chain of events that caused the failure? Did we see the looseness in the motor starter connection block? Did we see the expansion and contraction of the block, as the circuit heated and cooled, contributing to the looseness? Did we see the dirt fall in during installation that caused the resistance that led to the heating? If the failure is the result of a chain of events (and it usually is) what was the first link?
Step 3	Print out the existing task list and make a chart so that each failure mode described in Step 1 points to a task on the list. If you have a failure chain, see that items on the chain also point to tasks. One task can point to several failures. One failure can also point to several tasks.
Step 4	Ask: Are there failure modes that you are not looking for in the task list? Are there tasks in the task list looking at failure modes that never happen?
Step 5	Look at the cost of individual tasks compared with the cost of the failure modes you are avoiding.

Task list analysis is a microanalysis. In other words, every line of the task list is examined for suitability and every failure is examined to see if there is a task associated with it. It can be a painstaking process. Remember the payoffs: No unnecessary money spent on PMs and no unnecessary failures.

The sump pump:

Tasks	(some) Failure modes
Check bail, floats, rods, and switches. Touch each part.	Float is rotted off the arm, sump pump doesn't actuate when water fills sump.
Clean sump area	Float is fouled by debris in sump Switch fails.
Operate unit (fill sump with water from hose and observe that sump pump operates)	Circuit breaker off - Motor doesn't start.
* What if we add an additional task to remove the plug from the wall and clean it?	

** This additional task is not related to any failure mode. It also doesn't seem (on the surface) to be related to an E or S consequence. After discussion with the team we might drop this task.*

Your experienced team members would be able to think up tens of failures for simple equipment such as the sump pump and perhaps a hundred for a larger system. Each failure should be compared with the tasks to make sure it is "covered." As mentioned the failure mode should be "covered" if the cost of the task in the year is less than 70% of the cost of the failure in that year.

This task list takes 30 minutes or US$20 to accomplish per quarter. The cost is $80 per year plus the cost of the occasional short repair or corrective maintenance repair. We need to determine if use of our list is the appropriate strategy for the asset and for what it is protecting. Even in a house, the sump pump protects all the utilities, and anything stored at that level. In a factory or office building the stakes could be hundreds of thousands or millions of dollars.

Now let's take a look at an individual task. We have determined that each task is related to one or two failure modes.

Task	Time per year	Cost per year	Failure cost and estimate of frequency	Failure cost exposure per year
Clean sump area	10 minutes * 4 times per year = 40 min	$6.67 * 4 or $26.67 per year	Perhaps once in 20 yrs. debris would interfere with the proper action of the sump pump when the sump pump is needed. Cost when it does $4000	Failure exposure per year $200

For the sump pump we are spending $26.67 each year to avoid a cost of $200. This example is oversimplified to make the point, but it encompasses the steps necessary for task analysis.

After the task list has been in operation Mike Brown, President of New Standard Institute recommends a return look every year to see if: **1.** The procedure is still valid? Be sure the asset is still in service and no significant changes have taken place. **2.** All the performance limits still valid? **3.** Is the frequency still valid?

Some Advanced Concepts-PM at the Next Level

Did you ever think about the contradictions of good PM? The better the PM system, the more likely you would not have an adequate number of failures to analyze. Without the failures you lose sight of where your equipment is on the failure curve, or whether you are doing the right frequency of PM. Your crews start to lose their tear down skills because they are doing less and less heavy work on equipment that is completely broken down.

This is a real problem in certain highly critical industries. To get adequate numbers of failures the US aircraft industry aggregates maintenance data from all commercial airlines in the US into a single database. The FAA maintains records on maintenance work, rebuilds, and failures, in a computer center. The mission of this project is to alert the authorities, aircraft manufacturers, and the airlines, when a component starts to fail at a greater rate than predicted. It is also used forensically after a catastrophe.

The better you get, the harder it is to get even better. The only way around this problem is to increase the sophistication of the tools that you use to analyze the data.

In Chapter 9, in the introduction to the concept of frequency, we looked at using statistics to determine how often failures occur in your existing operation. Using your failure data we can construct curves that fairly represent how you currently care for equipment, how the equipment is used, and other factors. In this section we will increase the accuracy of that model.

Statistics analyzes the past and says that the future will look like the past. It has very powerful ways to turn a pool of failure data into a probability prediction for what will happen. Statistics works best as the population of equipment goes up. Decisions based on hundreds or thousands of fleet vehicles are more statistically valid than a study on 3 air compressors. This is one of the reasons why statistical analysis is not widely used in factories (but is more widely used in fleets).

The picture is incomplete. There is an area that statistics cannot easily look at. What does it mean if we say we have a failure with a mean of one per year? Generally it means that if we averaged all of the elapsed time between failures that average would be one year.

What would we do with that information? Using the statistics introduced in Chapter 9 we would pick a PM frequency to catch that event and inspect perhaps twice a year or even quarterly (depending on the SD and the intended or desired failure rate).

Lets look at a real example:

We are visiting an oil terminal that loads trucks for mostly local delivery. It has had ongoing failures in the PLC power supply module. Every few months the power supply would fail and the oil loading terminal would be forced to go into manual loading. Going manual significantly slowed down the operation, caused overtime costs, and excessive waits for the trucks. The trucks will wait for a while, but if the line gets too long they will leave and pick up their oil from your competitors. That volume is lost and cannot be recovered.

The problem was that the statistics showed clearly that with a 3-month MTBF, inspection should be at least monthly. The power supply got real hot before failure and it could quickly be inspected with an infrared gun (or tested by putting a finger on it.). The problem was that the elapsed time between initial heating and failure was only 24 hours. So for 89 days it was cool and then something happened and in one day it heated up and failed. It would have been a lucky inspector that caught and replaced the power supply hot and ready to fail.

The gap between good performance and failure here was narrow. In other electronic failures the gap might be only a minute or a second, or less.

At the other extreme is the propeller shaft of a steam ship. This 30-inch diameter shaft turns at about 15-20 RPM and runs from the engine to a large bearing and through a seawater seal. However, this shaft developed a bend. If you stood next to the shaft and held your finger just touching it, you could see the shaft move away and get closer as it turned. The engineer scheduled a dry dock stay a year ahead of time. It might be 2 or 3 years before this gradual increasing loss of function would cause a complete breakdown.

There is a curve that expresses this aspect of failure. The P-F curve (performance - failure curve) traces the performance of a component over time or usage. For a significant amount of the time covered the performance is stable (straight line parallel to the X-axis). At some point (CWP- Critical Wear Point) something happens and the curve starts to deteriorate. The something that happens might be microscopic and in itself completely undetectable by current technology. As the asset is used, the loss of function increases and the problem becomes easier to detect.

Performance Failure Curve- What does Inspection and PdM buy you?

Each of the points D, C, B, and A in the diagram represents a loss of function that is increasingly detectable. At point A the component functional failure can be detected by anyone (it's red hot or squealing loudly). After you pass point A on the curve the component will very soon have a complete loss of function or a breakdown.

The big message from the P-F curve is that PM and increasingly sophisticated PdM buy you time. You have time to buy spares, schedule down time, even time for training. The long ship propeller shaft breakdown time gave the engineer the ability to schedule a dry dock (which was much cheaper than breaking down anywhere on route).

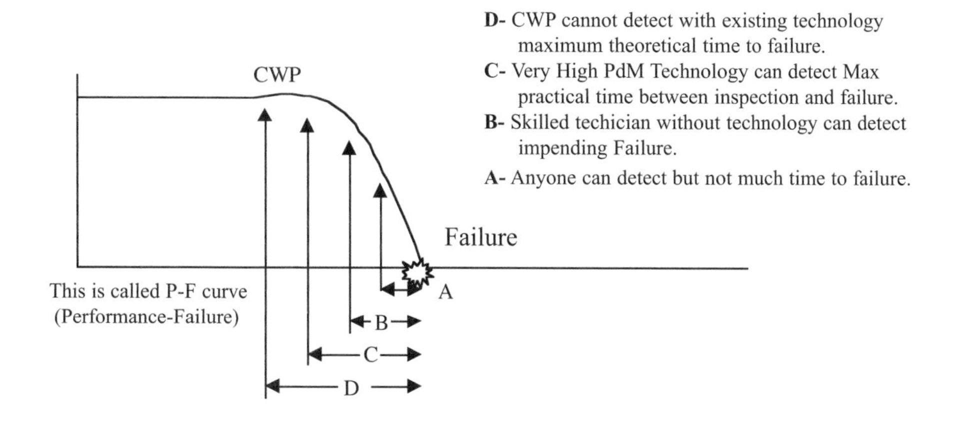

Using Statistical Analysis

PCR (Planned component replacement)

Using statistics, if we look at the failure interval we should be able to predict when the next failure will take place. In fact, if we had enough equipment of the same type in similar service we should be able to assign probabilities to the likelihood of failure. The field of statistics, when applied to failures of like equipment in like service can do just that. One outcome of this approach is PCR (planned component replacement)

PCR is an option on the PM task list. The novelty of this option is the elimination of failure because **components are removed and replaced after so many hours or cycles but before failure.** Depending on the sub-strategy some of the components are then returned for inspection, rebuilding, or remanufacturing, and others are discarded. The result of this strategy is controlled maintenance costs and low downtime. The strategy does not work when the new component experiences high initial `burn-in' type failures.

For example, fleets with time sensitive loads realized that breakdown costs with downtime are sufficiently high to justify PCR. It is standard procedure in some fleets to replace hoses, tires, belts, filters, and some hard components, well before failure on a scheduled basis. These soft items (belts, hoses) are called planned discard since there is no intention of using them elsewhere.

PCR is an expensive option. Even in the aircraft industry, significant effort has gone into improving reliability so that fewer components would be in the periodic rebuild program. According to John Moubray in RCM II, after an extensive RCM analysis the number of overhaul items (planned rebuild items) went from 339 on the Douglas DC 8 to just 7 items on the larger and more complex DC 10. Although the number has dropped dramatically, PCR is still an important tool to the maintenance professional.

PCR Made a difference

A few years ago I was flying to Dallas when the captain got on the speaker and told us that this was a historic flight. He said that this plane (a 727) was the first one that was made back in 1963 and was being retired after this flight (or actually sold to a freight company). After I got over the shock of flying on serial number 000001 (you are never supposed to buy the first of anything). I started to think about what that meant. It meant that the engines had been changed some 25 times, likewise the fuel pumps, hydraulic pumps, in fact everything had been changed. Companies were still improving components so that the plane had modern avionics and control systems. In short PCR made a new airplane every 10 or so years.

PCR is divided into two sub-strategies called planned discard (where you throw away the component) and planned rebuild (for rebuildable components like truck engines).

Planned discard is where a component is removed before failure and discarded. Common examples would include belts, filters, small bearings, inexpensive wear parts, etc. One fleet replaces hoses every two years during its major rebuild cycle to reduce the number of unscheduled hose failures.

Planned rebuild is for major components that are rebuildable such as engines, transmissions, gearboxes, pumps, compressors, etc. Components on aircraft are the best examples of this strategy. The items are removed after a fixed number of operating hours or take-off/landing cycles. They are sent to a certified rebuilder, brought back to specification, and returned to stock to be put on another aircraft.

In replacing the component before failure on a scheduled basis, PCR offers the following advantages:

1. The component doesn't fail. Some of the possibility of core damage is eliminated on planned rebuild parts. The value of the core is preserved. The core is the rebuildable item such as the alternator, pump, etc.

2. Replacement is scheduled so that you can avoid downtime and replace the component when the unit is not needed. Overtime from emergency repairs caused by untimely breakdowns, is thus avoided.

3. Expensive or rental tools can be made available on a scheduled basis to reduce conflicts and reduce costs.

4. Manufacturer's revisions, enhancements, and improvements can be incorporated more easily.

5. Rebuilds in controlled environments by specialists are always better than the same rebuilds ‘on the floor’ by general mechanics

6. Since it is scheduled, the rebuild can be used for training of newer technicians.

7. The PCR activity is great training for new or second-tier mechanics. All work is done on operating equipment so the mechanic gets to see how the equipment should look.

8. Spare components can be made available on a scheduled basis, which can minimize inventory (rather than waiting for breakdowns, which are known to clump together).

9. Since the component is replaced, breakdowns become infrequent, availability goes up, and conditions become more regular.

10. With a successful PCR plan, and to maximize the return from the investments, one assumption is that management will take the time to look at any failures and seek ways to avoid such failures in the future. Some options that can be looked into include better quality lubricants, improved skill in repairs, design review, and OEM specification changes.

Case Study: Comparing Breakdown with PCR Costs

A produce hauler currently uses P&M Truck leasing for full service leasing on their power units. They are interested in having your Central Garage provide maintenance services for their "refers" (refrigeration units mounted on trailers or rail cars). Alert readers will realize that this is a specialized model of the same type as introduced in Chapter 3 on economic analysis.

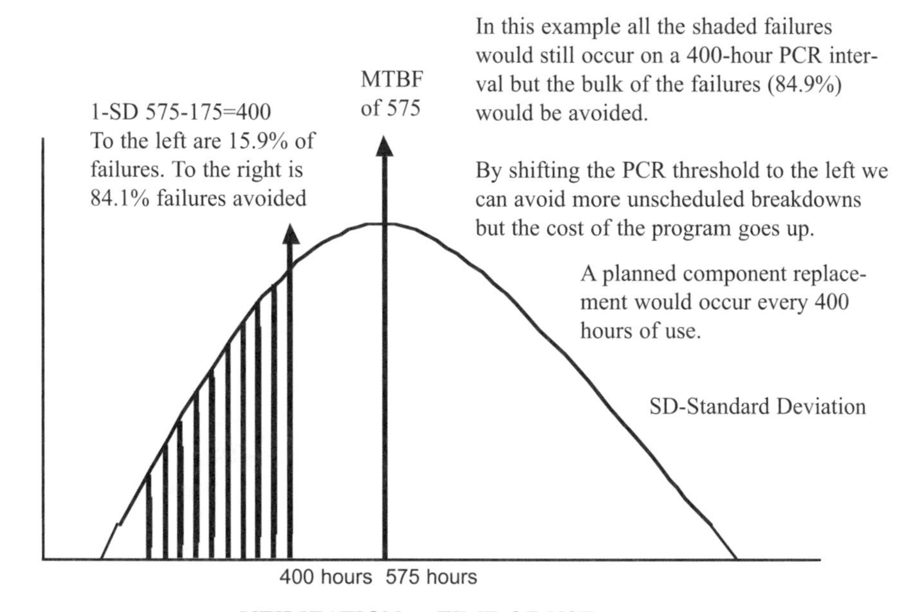

MTBF Curve with Standard Deviation

Facts for the Case:

50 TK (Thermo King) refer units mounted on Great Dane Refer Vans
Utilization 2500 hours per year, per refer
Belt failure rate averages 575 hours, SD 175 hr. (Mean-less 1 SD = 400 hrs)

Failure rate= 2500 hrs/ 575 hrs per failure = 4.35 per unit or 218 failures for fleet/ yr.
Cost per non-scheduled (emergency) failure $285.00
Cost per Scheduled replacement $85.00
Administrative cost per repair incident of any type $20.00

1. Cost for breakdown mode:

Failures * (Cost per failure + Cost of Admin.) = Total cost of Breakdown program
218 * ($285.00 + 20.00) = $66,490

2. Cost for PM using PCR: For greater accuracy, add the cost of PCR to the cost of the new breakdown rate to get the true cost of the program. Use 400 hours (1 SD from example) to pick-up 84.9% of failures:

(Utilization/PCR Interval)*units = PCR Incidents
(2500 hrs / 400 hrs)* 50 = 312.5 use 313

Failures * 15.9% = Emergency Incidents
218 * 15.9% = 34.66 use 35

Complete Cost of PCR

PCR Incidents * 313*	(Cost per incident + Admin. cost) ($85.00 + $20.00)	= PCR cost = $32,865
Emergency Incidents * 35*	(Emergency costs per incident + Admin.) ($285.00 + $20.00)	= Emergency Cost = $10,675
PCR Costs + $32,865 +	Emergency Costs = $10,675 =	Total Cost of PCR Program $43,540

Cost Breakdown: $66,490.00 **Cost PCR: $43,540.00**
Winner

Here the PCR alternative saves over $20,000.00. The mathematics of PCR can be tricky and the analysis can be time consuming. At least one firm offers software help. Their package, RELCODE helps the maintenance professional by digesting the MTBF and spitting out PCR frequencies. See Oliver-Group in the resource section.

Personnel Issues

Staffing the PM Effort

"A successful PM program is staffed with sufficient numbers of people whose analytical abilities far exceed those of the typical maintenance mechanic," (from August Kallmeyer, *Maintenance Management*). We want high skill and knowledge people with positive attitudes because they will be able to detect potentially damaging conditions before they actually damage the unit. Your best mechanic is not necessarily your best PM inspector.

What kind of person do we really want for this job?

Here's what I look for in a PM person. Six Attributes of a Great PM Inspector:

1. Can work alone without close supervision. The inspector has to be reliable because it is hard to verify that the work was done. Reliability has to be built-in because it is quite hard to add this attribute afterwards.

2. The inspector should also be the type of person who will fill out and complete the paperwork. The paperwork and subsequent write-ups for additional work need to be complete and accurate.

3. The PM inspector should know how to (and want to) review the unit history and the class history to indentify specific problems for that unit and for that class. Sometimes knowing about the most recent problems with a specifiic unit will indicate an area of weakness in the design. A great inspector will take an extra look where problems have been encountered in the past.

4. A mechanic is re-active in style. A PM inspector is **pro**-active in style. In other words, the inspector must be able to act on a prediction rather than **re**act to a situation. He/she is primarily a diagnostician, not necessarily a 'fixer.'

5. Because of the nature of the critical wear point, the more competent the inspector, the earlier the deficiency will be detected. Early detection of the problem will allow more time to plan and order materials, and will help prevent core damage.

6. PM inspectors should not be interrupted, and should be segregated (while they are in the PM role). PM is a mental process and needs extensive con-

centration.

PM is mostly boring. The PM mechanic is looking at healthy equipment, doing low skill cleaning, tightening, and lubricating, all to find the one unit that is wearing out. It is difficult to stay alert when it is the same thing day in and day out. The problem of PM is that without external reasons, without something else, PM will dull your mind. In this dulled state all kinds of mischief can occur.

Example 1

An automobile mechanic does a PM to a 2-month old car on a service request that the car would stall. The stalling problem was presumed to have been handled. In fact no one else had looked at this vehicle. The mechanic cleaned the terminals of the battery (which was a task in this PM routine). He didn't notice that the battery hold down clamp was gone. The battery had shifted in the holder so that the + terminal had shorted against the chassis. The shorted battery boiled over and spilled acid on the wiring harness, causing the harness insulation to be degraded to the point that exposed copper could be seen. If the owner hit a bump, the harness would shift and the car would stall. It takes something to miss all that damage. The service writer was in shock when shown what was passed as complete.

Example 2

A new wiper (person who does all the dirty jobs) joins a ship's engine room complement He is found to be intelligent and when the oiler jumps ship he is advanced on a temporary basis. He is shown a zerk fitting and a grease gun and is expected to 'go to town.' He sets up a grease schedule that lets him rush through the deck lubrication needs in a few hours and sunbathe on nice days, and he would do routes in the engine room when the weather was overcast or raining. After 17 days of pure sun and no visits below deck he started to rethink his strategy. But no one even asked.

How to insure the PMs are done as designed

One of the toughest problems to solve is how to insure that the inspector is actually doing the inspection on the task list. Horror stories about maintenance catastrophes frequently feature task lists that were signed off as completed but obviously not performed.

The challenge of leadership is to inspire the people in PM roles to want to do the tasks well. The inspector mentioned below (or for that matter mentioned anywhere in this section) can be a regular mechanic, operator, or helper (if appropriate) on a part time basis or a full time PM technician.

1. Does the inspector know how the PM activity fits into the overall scheme? Is it well known that PM impacts reliability, safety, costs, and output? You see the inspectors in nuclear power plants or in airlines, knowing full well the impact of missing a PM (and even then it happens).

2. Drag your top management down to the bowels of the facility and have

them address the maintenance crews about the criticality of PM and its impact on output or safety or whatever. You might have to write the speech. People attend to what they think management thinks is important. Let them hear it from the horse's mouth.

3. Present the job as important. If people feel that PM is stupid, boring, and low priority or fill-in work they are less likely to put themselves out.

4. One of the most important things you can do to insure the work is done is to let your PM mechanics themselves design elements of the system and tasks themselves. Have them trained in reliability, TPM, and general maintenance management and then let go of the reins.

5. One hole is lack of specific skills. Someone with a title of maintenance person, electrician or millwright should have the skill to perform the PM task. An individual might be lacking a specific item of skill or knowledge to effectively perform the task. Make absolutely sure that the PM people are fully trained. A test for PM certification might be appropriate.

6. Improve the relationship between the mechanic and the maintenance user. Where there is an operator such as a driver, machine operator, or a building contact person, instruct the mechanic to make personal contact. Some PM task lists include a task "talk to operator and determine if equipment, building, truck, etc. has operated normally since the last visit."

7. Make it easy to do tasks. Re-engineer equipment to simplify the tasks and route the people to minimize travel.

8. Simplify paperwork.

9. Improve accountability by mounting a sign-in sheet inside the door to the equipment. Be sure that the people who do the tasks sign a form and are included in discussions about the equipment. When people know they might be quizzed about an asset they are more likely to complete their PM tasks. When people know that an inquiry is conducted after a breakdown and the PM sheets are reviewed it motivates them to complete their tasks.

10. Make PM a game. One supervisor got a small amount of money and went to the local fast food restaurant and bought $0.50 gift certificates. Each week he hid 8, 3X5 cards (that said "see me") inside equipment on the PM list. He traded the cards for the certificates. He knew when a card wasn't found (PM wasn't done). His comment was "What people would do for $0.50 they wouldn't do for $17.50 an hour!"

11. PM professionals like new, better toys (sorry, better tools not toys). Technology has opened up the field for sophisticated, relatively low-cost PM tools. They might include $700 for a pen sized vibration monitor, $500 for a

cigarette pack sized infrared scanner, or $1500 for an ultrasonic detection headset and transducer. If appropriate to the size and type of equipment, these tools will motivate the troops and increase the probability that they will detect deterioration before failure.

12. In any repetitive job boredom sets in. Consider job rotation, reassignment, project work, office work like planning, design, and analysis to improve morale.

Get It Going Right

This entire chapter is one big list. This list recommends a series of projects, discussions, and meetings needed to install and support a typical preventive maintenance effort. Many of your organizations already have completed some of these steps. If the step is really complete, that is fine!

STEPS to Install a PM System

1. Create PM task force. The PM task force is a group that includes craftspeople (include the shop steward in union shops), a staff representative, data processing representative, and engineer. In some operations, a representative from operations is essential. This task force thrives where there is also a management champion (but they don't have to be directly on the team). Keep in mind that PM has four dimensions (engineering, economic, psychological, and management). Members of the taskforce should have expertise in one or more of these four areas and all four areas should be covered.

2. Decide on the goals of the task force. Set objectives. Begin to design the training program. Everyone on the taskforce must become an expert in PM (usually well beyond what they know to begin with). Do not start to design the program until the task force personnel are well trained in PM.

3. Pick a catchy name for the effort like PIE (profit improvement effort), DEEP (downtime elimination and education program, QIP (quality improvement program). Stay away from PM unless you can establish that PM does not have negative connotations for your people.

4. Sometime early in the process the taskforce should decide that PM is the appropriate strategy for this organization at this time. Prepare a complete economic Return-on-Investment ROI case study. Macro-analysis determines if the PM strategy is best for your organization. This economic analysis is essential to enroll top management in the process. This procedure is essential to get top management commitment and to get funds for the next steps. The case study should show the costs of the current operation, costs of the proposed operation, and the costs to get from point A to Point B.

5. Get training in computers for members of the task force if they are not computer literate. Include typing training. Get them access to computers and

any relevant organizational level networks or systems. At a minimum they should be able to use word processors, spread sheets, E-mail and presentation software. Much of the work of this task force can be shared by e-mail and can be designed in a Lotus Notes type environment. Intranet users can start a web site to inform members of the task force and eventually the rest of the organization. Of course training in PM should be well underway.

6. Part of the PM training is maintenance management training. Get generalized maintenance management training for the entire task force. This training will save time and effort by laying groundwork so that they can share a language and create a new vision of maintenance. There are many good teachers in every part of the world, so this training is money well spent. Some organizations build expertise by getting employees trained by a variety of the leading trainers and using them to build a unique vision for PM for their organization.

7. Identify the maintenance stakeholders (anyone impacted by how maintenance is conducted). Analyze the needs and concerns of the maintenance stakeholders. Use questionnaires, interviews, and common sense to determine each stakeholder groups' stake in the outcome. Look at each group and see how they contribute to the success of the organization. Tie in the plan to that outcome. This sequence is like showing how a reduction in breakdowns is demonstrated to reduce lost day accidents when making presentations to the risk management or safety people. Include production, administration, accounting, office workers, tenants, housekeeping, legal, risk management, warehousing, distribution, clients, etc. At least look at how your proposed changes will benefit each group. Consider drafting an impact statement for every group your change will impact.

8. Once the stakeholders are identified it is time to design and deliver the important "dog and pony shows." An example might be a (Power Point) slide presentation to various stakeholder groups that describes the program, lists the steps, and builds enthusiasm around the outcome. It should train the stakeholders in PM as it impacts them. Different shows can be developed for major stakeholders. Several (all) members of the taskforce should participate in the presentation. Enlist other groups to talk about their areas of specialty. Examples include production talks about the consequences of downtime, accounting about the impacts of costs, marketing could speak on the importance of this plan to smooth out delivery problems, etc. An especially effective tactic is to be sure that a member of the stakeholder group to whom you are speaking gives a part of the presentation. It is important to realize that no matter how good your PM plan is, and no matter how bad the existing situation, stakeholders have something to lose through the change and until they are convinced, nothing in their mind to gain.

9. Once buy-in has occurred, the core work of the project can begin. Design KPIs for the project. There should be several simple measures that will show the team (and other interested parties) how the project is going. With all KPIs, an up-to-date display of progress (or lack of it) is a powerful way to keep the project going. A display shows the data in an easy to understand format like gauges on a dashboard of a car (Joel Leonard's idea), a dial, or even a rocket going toward a goal. The key is to keep it up to date and accurate. Some KPIs to consider:

A. The first step is building the master files, so use a simple thermometer to show the percentage of assets entered and audited into the system. There are other master files, so perhaps include a measure for all of them or for critical ones only.

B. Once the files are built everyone must be trained. Percent of training completed could be another measure. A way to express this value might be an estimate of all training to be done, divided by training completed.

C. As areas are covered and work orders start to be issued, track hours reported by the system compared with payroll hours for the same group.

D. As PM tickets get issued, track numbers of PMs issued and completed each week.

10. Inventory and tag all equipment to be considered for PM. Compile and review your list of equipment. If you have an effective CMMS then this step should be complete. If so, audit the master lists of equipment. Pass the list out to operations for verification. This list is a starting point for the PM program. Inquire if lists exist in plant engineering or accounting.

Be sure to Look at:

Access items such as doors, windows, hatches
ADA requirements (disabled access)
Boilers
Chemical storage
Clean rooms
Communication systems, raceways
Compressors and air delivery systems, vacuum systems
Computer rooms, shop floor computers
Control systems (like PLC's, MAP systems)
Drain systems, environmentally secure means of disposal
Elevators, people movers
Electrical items (major), electrical distribution systems, transformers, sub-stations
Environmental systems (scrubbers, separators, filters)
Environmental inspection (asbestos encapsulation integrity)
Food service equipment, kitchen equipment, laundry equipment

Exterior finishes, accessories, roofing, roof catwalks, equipment attached to roof, openings
Generators, co-generation facilities, power houses
Grounds, pavement, sidewalks, parking areas
HVAC components (heating, ventilation, air conditioning) exhaust systems
Interior finish, lighting
Legal liability inspections such as fire systems, elevators (use contractors?)
Mobile equipment, trucks, trains, cranes, ships, cars, pick-up trucks, turf equipment
Plumbing items (major), pumps, piping systems, rest rooms
Production equipment, process equipment
Quality inspections, certifications, ISO 9000 requirements
Rack systems, automated conveyers, and storage/retrieval systems
Safety/security systems: fire alarm, fire extinguisher, smoke detectors, and security systems, physical structure of building
Swimming pools, settlement ponds, water intakes
Tanks (both underground and above ground), related piping systems, chemical reactors
Trash compactors, trash-handling systems, and recycling systems
Waste, HAZMAT handling systems

Identify the equipment that is important for the sucess of your business. This becomes high priority equipment.

11. Select a system (CMMS is the usual choice) to store information about equipment, select forms for PM generated MWO and Check-off sheets. Again, if you already use a CMMS then the choice is complete. The challenge is to build the task forces' expertise in the PM module for your particular system. All the CMMS are slightly different. These differences might seem trivial on the surface but poor choices could make the job much harder. Expertise with the specifics of your system is essential. Try to attend a user meeting for the CMMS if there is one.

12. Design first drafts of the measures or benchmarks called KPIs (Key Performance Indicators) to be used to evaluate the ongoing PM system's performance. At different stages in the projects, different KPIs are needed to move the project along. These measures will be revised as the process goes on. The KPIs designed in #9 can be morphed into the KPIs with little extra effort.

13. Take a complete look at your business process. Chart the steps necessary to get PM done. Consider changing the business processes to speed them up and reduce the time needed. As the new process is designed, begin to draft

SOP (standard operating procedures) for the PM system. This document also will need to be revised many times over the first year.

14. Have task force members or shop personnel complete data entry or preparation of equipment record cards (if not already complete from the CMMS installation). Rotate the data entry job so that many (everyone?) in the department has experience collecting, adding, and auditing data. Widely held experience in correcting mistakes in the database is necessary before you go on line. It is essential to build a critical mass of expertise in the system. If this is a CMMS installation there are two levels to the effort. One is to collect a complete list of all equipment. The second level is to collect all the nameplate data and add that to the files. These two different tasks can be done sequentially or at the same time. If the CMMS is operational build the PM module (but the rules above still apply). Enter the task lists and frequencies and relate them to individual equipment. Enter parts data, part kits, tools required, lock-out/tag-out steps, PPE, PM task steps, and other planning data to make the PM tickets instantly useful.

15. Be sure to replace hours invested in the system. Bootstrapping the PM system will hamper the effort and make it take longer. Consider using contractors and some overtime to replace the hours lost on the floor by the people doing the data entry.

16. Fight the tendency to use the CMMS vendor to build the details of the PM system. The vendor can be directly involved (if you feel you need the expertise) but only as an advisor. Ideally when needed the vendor should be hired as advisors, auditors, cheerleaders, and councilors but not on the playing field as a team member.

17. The ongoing daily audits of all task list and support data typed into the system constitute another essential task. Have someone who is highly skilled review all data going into the system.

18. Select people to be inspectors. Allow their input into the next steps. Consider using inspectors to help set up the system. Consider letting this team take on the steps of creating the PM system. Certainly the system will be handed to this group after the task force has been dissolved.

19. Get training in RCM (reliability centered maintenance) or PMO (PM Optimization) and failure analysis for key personnel. This training will help them and the program immeasurably, showing them how to root out useless tasks and include important tasks on hidden functions.

20. Determine which units will be under PM and which units will be left to breakdown (BNF or bust and fix). Remember that there is a real cost associated with including any item in the PM program. If, for example, you spend

time on PM's for inappropriate equipment you will have less time for the essential equipment. Costs to include in PM Program are:

Cost of Inclusion = Cost per PM * Number of PM per year.

To decide which units to include in the PM system, apply the following tests to each item:

A. Would failure endanger the health or safety of employees, the public or the environment?
B. Is the inspection required by law, insurance companies, or your own risk managers.
C. Is the equipment critical (vital to the success of the entire enterprise)?
D. Would failure stop production, distribution of products, or complete use of the facility?
E. Is the equipment the link between two critical processes?
F. Is the equipment a necessary sensor, measuring device, or safety protec tion component?
G. Is the equipment one of a kind?
H. Is the capital investment high?
I. Is spare equipment available?
J. Can the load be easily shifted to other units, or work groups?
K. Does the normal life expectancy of the equipment without PM exceed the operating needs? If so, PM may be a waste of money.
L. Is the cost of PM greater than the costs of breakdown and downtime? Is the cost to get to (to view or to measure) the critical parts prohibitively expensive?
M. Is the equipment in such bad shape that PM wouldn't help? Would it pay to retire or rebuild the equipment instead of PM?

21. Once it has been decided that an asset (machine, unit etc.) is to be included on the PM system (the macro analysis has been completed) the micro-analysis begins. A close examination of the failure history, collected information from the OEM, and the accumulated experience of your maintenance and operations team are now to be focused on that machine. The next several items on this list deal with aspects of microanalysis.

22. Schedule modernization on units requiring it. Investigate the possibility of retiring bad units if possible. A bad unit left on the system will demoralize the most dedicated inspectors.

23. Select the PM clock or measurement system you will use (days, utilization, energy, add-oil). A clock is designed to indicate wear on an asset. Clocks on items in regular use or subject to weather are usually expressed in days. An irregularly used asset might be better tracked by usage hours or output tons of steel, cases of cola, etc. Some items such as construction equipment

are best tracked by gallons of diesel fuel consumed because hour meters are frequently broken.

24. Decide what Predictive Maintenance technology you will incorporate. Train inspectors in techniques. Even better, provide the information and a budget to the task force and let them pick the technology. Most equipment should be rented to try it out before buying. Inexpensive training is available from most vendors and distributors

25. Set-up task lists for different levels of PM and different classes of equipment. Factor in your specific operating conditions, skill levels, operators experience, etc. Consider all the strategy including unit based, string, route maintenance, and future benefit as well as non-interruptive /interruptive. Consider what strategy to use to schedule PMs and what to do if the date slips. Be sure to include a review of the actual failure history when designing the task list. It is great to design for possible failure modes but it is essential to design for failures that have actually occurred.

26. Start a program of public relations. Sift through your data and find statistics that indicate success stories. Identify those stories and write them up as powerful narratives. Publicize your successes. It is okay to publish stories from you industry or from other industries where similar equipment is involved. One idea is to collect PM stories or maintenance catastrophes (that you can show would not have happened with PM) from various sources and circulate a different one every month.

27. Document all PM tasks. Categorize the PM tasks by source (recommended by Ron Moore, of RM Group). Categories might include regulatory, calibration, manufacturer's warranty, experience, insurance company, quality, etc. This documentation will be a great aid when you look back to see which ones to eliminate or change.

28. Provide the PM inspector with the following to perform the tasks:

A. Task list (usually printed on a work order) with space for readings, reports, observations
B. Drawings, performance specifications, pictures where appropriate
C. Access to unit history files and trouble reports.
D. Equipment manual(s)
E. Standard tools and materials for short repairs.
F. Consider having a cart designed for the PM's and common short repairs
G. Any specialized tools or gauges to perform inspection
H. Standardized PM parts kits
I. Forms to write-up longer jobs to be submitted to maintenance dispatcher.
J. Log type sheets to record short repairs or (if your system will allow it)

short repairs may be added to the bottom to the PM sheet and entered into the system. It is important to capture short repairs in the CMMS (if possible).

29. Assign work standards to the task lists for scheduling purposes. Observe some jobs to get an idea of timing. Let some mechanics time themselves and challenge them to re-engineer the tasks to cut PM time. Remember that time spent on PM does not itself add value to your process. The goal always is to minimize PM time while getting the task done correctly.

30. Engineer all the tasks. Challenge yourself to simplify, speed-up, eliminate, and combine tasks. Improve tooling and ergonomics of each task. Always look toward enhancing the worker's ability to do short repairs after the PM is complete.

31. Determine frequencies for the task lists based on clocks chosen. Select parameters for the different task lists.

32. Implement system, load schedule, and balance hours. Extend schedule for 52 weeks. Balance to crew availability. Schedule December and August lightly or not at all. Allow catch-up weeks throughout the year.

33. Plan to have a periodic meeting with the task force (as it is now constituted) to evaluate the on-going use of the system

The Future of P/PM

One point that is commonly missed is that PM is a way station to the ultimate goal of maintainability improvement. Remember the Holy Grail chart in Chapter One. PM can be an expensive option because it requires constant inputs of labor, materials, and downtime. The ultimate goal of maintenance is high reliability without the inputs.

Let's predict the future.

The question that will drive maintenance is; what does management really want from the maintenance effort? The answer to this question will vary from organization to organization. To confuse the issue, in most cases there is a public agenda that is discussed and a secret agenda that is not.

The public face of organization talks about the missions and goals and support for the workers, communities and customers. Those aspects are important, but we have seen a consistent subrogation to short term profits or short-term financial goals. The not-so-secret agenda is to get rid of all efforts that don't (in a very limited sense) make products. The secret agenda of top management is to get rid of the maintenance effort.

As maintenance professionals we may not like this future. As knowledgeable maintenance people we might know this is impossible. Maintenance people have an advantage over other professions. We spend our lives in the real world. Every day we are forced to deal with realities that we don't like or don't agree with but that must be dealt with. In other words, your opinion about the busted machine doesn't make a whole lot of difference!

Less Is More! This phrase will be the rallying cry of smart maintenance organizations. They will find more ways to cover the operation with fewer resources each year. Great maintenance departments will study maintenance activity and develop ways to do more and more with less and less. Wasted motion, material and mental effort will be attacked as the enemy.

This intense study will conclude that the solution is the design and redesign of systems that are intrinsically more reliable. Better design will result in greater and greater levels of reliability with lower levels of effort. To achieve this dream maintenance activity will be viewed in a special way.

One change is the push for permanent solutions. For maintenance departments to have the same failures as last year will no longer work. Each year we will be learning things that make some types of failure modes obsolete.

There will be a sustained push to proactive activity. This drive will identify items needing work well before failure. When a failure does (infrequently) occur there will be a reflective attitude to look at the root cause of the problem. In the best organizations there will be less finger pointing and a greater drive to understand what actually happened. Once the failure is understood we will redesign the system until the defect is worked out. The system will then be intrinsically more reliable.

Perhaps this leap is too far, but it seems that in the old days management were cowards. They would hide behind ignorance when it came to deferred maintenance. Management would often avoid funding deferred projects and then cry foul when the item failed (sometimes to horrific consequences). They wouldn't "put their money where their mouth was."

Well in this future- NO MORE. Management will be held responsible for what happens on their watch.

One of the symptoms of a new future was demonstrated by the bursting of the Internet investment bubble at the end of the 1990's and just into 2000. When we were inside the bubble amateurs ruled. No one had been in this environment before so it was assumed, by almost everyone, that experience and professionalism didn't count anymore. It was assumed that business cycles were transcended. These attitudes extended up to the top managers and down to the plant level. Really stupid business decisions were made to pump up the IPO prices that were to the detriment of the business as a business.

Through fits and starts we are settling on a future dedicated to substance rather than fluff. Things have changed. Now there is no room for amateurs. The survivors of this era will be the ones who focus on their business rather than boosting their profits through financial manipulation.

At the core of all business is a group of dedicated professionals who understand what it takes to turn crude oil into gasoline and what it takes to maintain the equipment to perform that task 7 days a week, year in and year out. Organizations will either honor this expertise or they will fail.

People spend the first few years of their career learning what they think are the rules of maintenance management. In the professions of accounting, law, or even engineering, the rules are taught in college and the practitioners learn to apply what they learned to the situations they faced in the work place. Well there is bad news. You will look long and hard and the rules will vary by company, by industry, and even by maintenance manager. In other words there are no easy accessible rules in maintenance. You have to learn the hard way and make up your own rules!

One trend that will accelerate is the demand for hard numbers from the maintenance department. Management has always been wishy washy about requiring work orders and other maintenance record keeping to be accurate and complete. To properly analyze any alternative we will need access to good numbers that are held by both management and workers to be accurate. Hard numbers will be king.

One of the uses of accurate numbers is the ability to compare one operation with another. This comparison is called benchmarking. There was a flurry of benchmarking studies in the 1990s (benchmarking became the management flavor of the month!). Benchmarking seemed to die away. In the future there will develop a serious use of benchmarking designed to shed light on areas of the business where we have just been getting by, and we have not been improving. Benchmarking, in its best form, can provide a sobering view of our business efforts. There is always someone better at some aspect of maintenance, no matter how good you are.

One other reason to measure maintenance KPIs (key performance indicators) accurately is to see if improvements really improve the department's performance. There are so many promises and so little follow up to see if promises were met. Proper benchmarking makes the promises public and the yardstick public at the same time.

In this future we are weaving we have to ask the question: are we getting better at maintaining the asset base we are responsible for? For an individual the question becomes: is my knowledge increasing every year or have I become stagnant? In the organization sense: can we maintain these assets for less resources every year? If not why not?

Part of the new mission requires willingness to run controlled experiments. In the future the maintenance department will command a significant research and development budget (something to look forward to). The reason will not be altruistic. One of the largest uncontrolled expenses in industry is maintenance. Investments in better tools, techniques, and materials can generate significant returns. Sober management wants to invest where the money can be made!

Always: Focus on service to the customer or focus on adding value to the customer. In the future a great maintenance department will listen to and talk to their customers. In fact they will be communications animals. Once the immense expertise and can-do attitude of the maintenance department is available to the organization through communication everyone will want maintenance input to improve their projects. PM choices will be made between maintenance experts and newly expert maintenance customers. Once maintenance understands the customer's true needs and the customer understands maintenance's constraints the decisions will improve by orders of magnitude. One way to serve the customer is to let them play with us. A powerful and slow growing trend is customer participation in maintenance (with training!)

We are the world's experts in tool use. We strive for the best, most efficient tool for every job. In this future we are discussing there will be a willingness to use sophisticated tools of statistics, finance, and accounting in maintenance analysis. We will uncover and learn to use every tool that can shed light on the maintenance reality. Another way to say this is: we will be proponents of Analysis Driven Maintenance (ADM) not the more traditional Seat of the Pants Driven Maintenance (SPDM-which has been popular until now).

It all boils down to people. Every problem is a people problem. In fact people are your only asset. Cross training (also known as multi-skilling) will be the order of the day. The maintenance department will become a kind of school with continual training. The maintenance department will take the lead in training the entire organization in lessons learned while working directly with equipment.

Every opportunity is a people opportunity. How people are used and how they feel about how they are used is key. Attachment to the people rather than to technology or computer systems will be increasingly paramount in future business. As the systems get smarter and stronger they allow more input from users and more flexibility that will lead to better data in and better decisions out.

Today is a time of layoffs whenever there is a glitch in the quarterly results. This short term thinking leads to organizations that can imagine a future more than 1 or 2 quarters away. Be sure your primary attachment is to your people so that every other option is looked at before layoffs (W.E. Deming says "drive out fear"). Maintenance has been traditionally pretty good with people issues. You don't have to spend very much time with maintenance professionals to realize how much skill is needed to be effective.

Maintenance has always used the team concept. What will be added is the fading of traditional departmental barriers. Information is everywhere in the company. Maintenance only has a small piece of the puzzle. The best decisions in any single domain require information from around the entire organization. As we reduce the interdepartmental walls we also learn to share our information and knowledge.

The Goal: Powerfully self motivated workforce and excellent execution of maintenance.

Important Question: Is your organization ready for the future? If not it will be eaten by one who is!

Common Mistakes

•Blindness as to the pattern in the target operation. In other words you hide from the ultimate reality of your operation. Acting as if what you wished was true was actually true.

•Hope for instant pudding (Deming), How much organizational patience does your firm have? Can it support things beyond a budget period?

•Ignoring the workers (all wealth flows from their hands)

•Ignoring the middle management (they can make it or break it)

•Thinking that this is only a maintenance project

Success Timetable

•Design a 6 year plan

•Revise it annually

•Look for low hanging fruit initially to generate ROI but don't expect returns until 18 months have passed

•Cross pollinate where ever possible

Make your own rules,

Good Luck
Joel Levitt

Appendix A: Tasks

Disclaimer: The author and the publisher take no responsibility for the completeness or accuracy of any task list contained herein. They are examples only. It is your responsibility to insure completeness. Also, before you use these lists or elements of these lists, you must add the proper safety, personnel protection, and environmental protection steps for your particular equipment and operating environment. It is your responsibility to evaluate the individual risks of your equipment, facilities, and environment and add tasks accordingly.

Contractors and service bureaus are excellent sources of task lists. One of the best known is HSB Reliability. Some CMMS organizations also accumulate task lists from years of working with clients. Some examples from eMaint are from this category. A discussion of their approach follows. Another, which the author worked with, was TPM Service Bureau.

TPM was a PM service bureau concentrating on buildings and facilities, that was started by HRM Associates and then operated by Titan Software, and eventually by Four Rivers Software. When you signed up for their PM service bureau an engineer would visit and make a list of your major assets, they called them MWIs (Maintenance Worthy Items).

The engineer carried a book of generic lists. During the survey he or she would photocopy these generic task lists for the items in that building, make a few quick customizations and produce a complete PM system. It was an extremely quick and painless process that worked well for smaller operations. After set-up, all data and reports were sent back and forth by mail. Completed PM tickets went to the service bureau and new PM tickets and reports flowed back to the customer.

The generic lists were very useful for maintenance situations such as apartment buildings and other tenant occupied buildings, where there were not many maintenance workers and not a deep knowledge in maintenance management. These generic PM lists were a good starting point for the average operation. It was up to the engineer or eventually on-site personnel to make appropriate modifications. One pitfall that arose was when the list contained something that the MWI didn't, such as a fuel pump on a natural gas boiler. When a new piece of equipment was encountered the engineer gathered the manufacturers' list, added in any of their own experience and came up with a new generic standard.

Many of the lists were simple and obvious. These lists helped the owner keep the asset in good condition without having to manage the process personally.

In the next few lists, see if you are given enough information within the list to perform the task. All of the checks in the first lists are visual checks looking for the integrity of the item. The question is how skilled do the inspectors have to be to perform the task?

The bulk of the task lists were provided by eMaint, a CMMS company in New Jersey (contact information in Resource section). These task lists are part of a library available to users of their CMMS, and are very good. The only things that are missing are frequencies (in most lists)) and the make-up of the different tool kits.

The other big opportunity is to look for what the inspector would have to carry to be able to do most short repairs that the inspection would turn up. In the interior he or she would have to be stocked with a ladder, all kinds of lamps, rags, vacuum, paint, electrical outlets, switches, cleaning supplies, etc.

The exterior lists call for some other materials and some decisions. Do you want glass to be replaced as a short repair, do you want (a few inches of) pavement sealed, etc. In other words it is not only the PM but the short repairs that are important.

Critical questions to ask include what failure modes are being looked for, how much does this task list cost to execute? In some examples the failure mode is someone looking at the building and walking away because they don't like what they see. The list can be modified to suit your building, factory, or even bus station. Note that members of housekeeping can use this same list; elements can be done by security personnel too.

Again, other groups or contractors can do elements of this list. Failures in buildings tend to take longer and produce more symptoms than in production environments. Water is usually the enemy. It does its damage over many years. The cost of the corrective maintenance increases dramatically when the facility is left to deteriorate.

Once we get into interior mechanical items, safety becomes a big issue. In addition to safety the skill requirement goes up. For example what is excessive noise in a bearing? The inspector would have to be a grade up to perform these tasks, in which there is a lot of implied knowledge. This requirement is not necessarily a problem as long as you can assure that everyone doing the tasks is qualified. Consider testing to be sure everyone is qualified.

Governmental agencies sometimes supply useful task lists

In the US the Department of Housing and Urban Development (HUD) has a program to subsidize housing costs for poor Americans. The program, known as Section 8, is administered by local agencies throughout the country. The agencies are all required to inspect apartments under Section 8 annually and determine if they meet the Housing Quality Standard. Their inspections not only include maintenance items but the adequacy of the facilities.

The entire inspection is on a PASS,-FAIL basis. Any FAIL scores have to be repaired within 30 days. Clear and present hazards must be repaired within 72 hours.

The second thing that you notice with these generic lists are that all lubricants and quantities are not spelled out, but they must be. Of course if work is required on any equipment, a person certified on that equipment is necessary (CFC license, boiler license, etc)

Loftin is a contractor servicing generator sets and transfer switches for their headquarters in Phoenix, AZ. They suggest that the generator set and transfer switches are one of the most neglected areas in large buildings. Loftin can be found in the resource section.

Some PMs have extensive special instructions such as this one from eMaint. EMaint is a CMMS company (in resource section) and they include over 80 task lists in a library with the software they supply. Many of them have been included here.

The system allows you to cut and paste from the PM library to quickly build the PMs for your individual buildings and factories. The PMs include safety, environmental security, and reminders to Read the (expletive deleted) Manual! The Maint library does not include frequencies.

Generic Interior Inspection

M	Check condition of walls, floor, and ceiling; report on any damage	
M	Check condition of switches, outlets and other electrical items.	Short repair-switches, outlets, covers
M	Replace all burnt out lamps, wipe off diffusers	Rags, sponges what lamps?
M	Replace all burnt out lamps in exit signs, wipe off exit signs	Lamps
M	Check condition of all doors and locks in the area	
M	Check condition of all windows and locks in the area	
M	Check condition of railings throughout interior.	
M	Verify count of fire extinguishers and verify dial is in GREEN area	How many, where?
M	Verify fire extinguishers have not been discharged and are not dented or damaged	
M	Check condition of fire pull stations, bells, heat sensors, and smoke detectors. Wipe dust off sensors; pull stations, and vacuum smoke detectors.	Vacuum cleaner, rags

Generic Exterior and Roof Inspection

M Check condition of paint, siding, stucco, siding (Carry paint?)
M Check for broken windows and doors
M Replace any burnt out exterior lamps (Carry lamps?)
M Check condition of all railings (Wrench to tighten)
M Check any exterior electrical connections and boxes
M Check for plants growing on building or into foundation Pull out if appropriate
Q Clean roof -Use care when working in high places
Trash bags, fall protection such as "use safety line with belt" if necessary?
Q Clean roof drains and gutters. Test drains and/or downspouts by flushing with water. Where applicable, examine strainers in drains and/or screens over gutters.
Q If downspouts have heaters, test operation and correct deficiencies
Q Inspect roof (at least perform inspection prior to heating and cooling seasons.) Consult manufacturer's/builder's information for type of roof membrane.
Q Check condition of antennae and wires
Q Inspect gutters for adequate anchors and tighten if necessary.
Q Inspect stacks and all penetrations through membrane
Q Remove any plant life growing on the roof, following approved methods. Do not allow roots to penetrate roof.
Q Clean up and remove all debris from work area.

Generic Roof Inspection

Inspect roof (perform at least one inspection prior to each heating and cooling season.)) Consult manufacturer's/builder's information for type of roof membrane. Use care when working in high places. Use fall protection such as safety line with belt if necessary.

1. General Appearance- check for cans, bottles, leaves, rags, and equipment that may have been left from job on or near the roof. Dispose of appropriately.
2. Water Tightness- check for presence of leaks during long-continued rain, leaks occurring every rain etc.
3. Check exposed nails that have worked loose from seams, shingles and flashings.
4. Check for wrinkles, bubbles, buckles, and sponginess on built up roofing.
5. Check exposure of bituminous coating due to loose or missing gravel or slag.
6. Check shingles for cracking, loss of coating, brittleness, and edge curl.
7. Check seams on built up roofing.
8. On wood shingles, check for cracks, looseness and rotting.
9. Check for water ponding.
10. Check all flashing for wind damage, loss of bituminous coating, loose seams and edges, damaged caulking and curling, and exposed edges. Check flashing fasteners for loose ness and deterioration.
11. Check all metal gravel stops for damage and deterioration.
12. Inspect all pitch pockets for cracking, proper filling, flashing, and metal damage.
13. Check lead sleeves on roof vents for deterioration.
14. Check inverted roof systems for fungus growth in between and under insulating panels.

Generic Grounds Inspection

W Check grounds for broken glass and debris (Trash bags)
W Check condition of sidewalk
W Check condition of driveway and parking area
W Clean storm water drains
W Check condition of lawn and plantings
W Verify no tree limbs are about to fall
W Check cleanliness around dumpsters
W Check condition of fencing
W Check mailbox area

Generic Task list for an apartment HVAC system

M Clean air intake
M Change air filter
Q Inspect condition of gas piping, burners, valves
Q Check blower motor in operation for excessive noise or vibration
Q Clean motor and ductwork
Q Check condensate drain pan for proper drainage
Q Check flexible duct connectors
M Secure loose guards and panels
M Check condition of electrical hardware and connections
M Check safety controls and equipment
Q Check for proper operation of interior unit
M During the cooling season check condenser motor bearings for excessive noise or vibration
M During the cooling season clean condenser air intake, discharge, and coil as required.
Q During the cooling season check condition of electrical hardware connections
Q During the cooling season check condition of refrigerant piping and insulation
M During the cooling season secure loose guards or access panels
Q During the cooling season check operation of exterior unit

Section 8 Apartment Inspection

In a kitchen, for example the inspector would check for:

Kitchen area present	Floor condition
Electricity on	Stove with oven
Electrical hazards	Refrigerator operational
Adequate security	Sink operational
Window condition	Food prep and storage space
Ceiling condition	Lead paint

In the apartment bedrooms, the inspection is pretty simple

Adequate illumination	Floor condition
Electricity on	Window condition
Electrical hazards	Ceiling condition
Potential hazards	Lead paint

The bathroom:

Flush toilet operates	Floor condition
Fixed wash basin operates	Window condition
Tub or shower operates	Adequate ventilation
Adequate illumination	Ceiling condition
Electricity on	Lead paint
Electrical hazards	Potential hazards

For everything else:

Condition of foundation	Floor condition
Condition of stairs, railings, and porches	Window condition
Condition of roof and gutters	Adequate ventilation
Condition of chimney	Ceiling condition
Adequacy of heating equipment	Lead paint on exterior surfaces
Adequacy and safety of water heater	Other potential hazards
Approvable water supply	Garbage and debris accumulation
Adequate and safe plumbing	Other electrical hazards
Sewer connection	Access to unit
Interior air quality	Evidence of insect infestation
Smoke detectors	Trash disposal
Fire extinguishers	Fire ladder

Generic Boiler

M Check all relief valves for free operation and leakage
M Check operate all water, gas, and fuel, valves for free operation and leakage
Q Check condition of insulation on boiler and stack
Q Check all manifolds for leakage
M Check all water, gas, and fuel piping for leakage
M Clean, lubricate, and assure free movement of all linkages
M Check operation of all motorized valves
A (If Oil) change fuel filter
A (If run on oil last year) Check condition of V-belts
W Check fan motor bearings for vibration and noise
Q Lubricate fan motor
Q Clean motors
W Check condition of electrical hardware and connections
W Check operation of all safety and automatic controls, including limit and flame safeguard controls
M Check operation of low water cut-off
Q Check condition of air separator at ceiling
Q Check condition of paint
W Secure loose guards and access panels
W Check operation of boiler, witness startup and shutdown, check for excess smoke
Q Check for operation of draft control on wall
S Perform efficiency test
W Check condition of boiler temperature and pressure gauges, record readings
W Check condition of stack temperature gauge, record readings
Q Treat boiler water (or assure contractor has)
S Perform bi-annual cleaning of the water side surfaces by flushing with water
A Perform internal and external inspections

Generic Domestic Hot Water Heater Task List

Q Check all gas connections for leaks
Q Check condition and operation of gas burners, gas valve
Q Check condition of insulation
Q Check water lines for leaks
Q Operate relief valve check for leaking, free operation
Q Blow-off water from bottom of tank
Q Inspect flue for obstruction and correct operation
Q Secure loose guards and panels
Q Verify temperature controls operate
Q Check condition of paint
Q Verify operation of hot water heater

Generic Roof Exhaust Fans

Q	Check motor bearings for excess noise or vibration
Q	Clean motor
Q	Check V-belts, replace or adjust
Q	Assure pulley set screws are light
Q	Check fan bearings for excess noise or vibration
Q	Clean air intake and discharge
Q	Check and verify operation of safety controls and equipment
Q	Check fan operation and local stop switch
Q	Secure loose guards or access panels
Q	Check operation of exhaust fan

Generic Pump PM Task list

W	Check motor bearings while in operation for excess noise or vibration
Q	Clean Motor
Q	Check condition of coupling between pump and motor
Q	Lubricate pump bearings
W	Check pump bearings while in operation for excess noise or vibration
W	Check pump seals for leakage
W	Check all piping, valves for leakage
W	Check condition of suction and discharge lines
Q	Check and record readings on temperature gauge
W	Check condition of insulation
Q	Check condition of electrical hardware and connections
Q	Check and verify all safety controls and equipment
Q	Check condition of paint
W	Secure loose guards and access panels
W	Check operation of unit

Generic Trash Compactor

- **W** Check motor bearings on pump for excessive nose or vibration
- **M** Clean motor and pump
- **W** Clean compactor
- **W** Check oil in reservoir fill as required
- **M** Clean vent breather on reservoir
- **W** Check condition of hydraulic hoses
- **W** Check inside bin and chute for obstruction
- **W** Check condition of compactor, fasteners, floor mounting
- **M** Check condition of electrical hardware and connections
- **W** Secure any loose doors, guards, or access panels
- **W** Check operation of electric eye
- **W** Check safety controls and equipment
- **W** Check for any fire danger
- **Q** Check condition of paint
- **W** Check for proper operation of pump and compactor
- **W** Check rams for free operation
- **M** Assure readiness of sprinkler system

Inspection for (generic) Generator Set

- **M** Inspection of cooling system fan, fan blades, remote cooling fan motor.
- **M** Inspection of all cooling system hoses, and adjustment of hose clamps if necessary.
- **M** Inspection of engine belts and belt tensions, with adjustment if necessary.
- **M** Inspect engine block heater for proper operation, temperature, and flow.
- **M** Inspect and clean generator controller and area, if required.
- **M** Inspect and clean gauges for proper operation, and adjust if needed.
- **M** Check shut down functions, including emergency stop for proper operation.
- **M** Inspect Automatic Transfer Switch for proper operation, with or without load.
- **M** Check time delays in Automatic Transfer Switch for settings.
- **M** Check and adjust exercise clock timer in Automatic Transfer Switch.
- **M** Verify proper operation of Remote Annunciator panel.
- **M** Check all bulbs in controller for proper operation.
- **M** Inspect and test both engine battery charging alternator, and the system battery charger, and adjust if necessary.
- **M** Start and run generator set to verify proper operation of unit.
- **M** Check and adjust all gauges.
- **M** Check anti-freeze / coolant level, and adjust if necessary.
- **M** Inspect generator for oil, fuel, and coolant leaks.
- **M** Inspect exhaust system and silencer for leaks, cracks, and deterioration.
- **M** Drain moisture for exhaust piping (if equipped).
- **M** Check batteries for water level, level of charge, and corrosion on terminals.
- **M** Check fuel system, including day tank or transfer tank (if equipped).
- **Y** Change lubricating oil and filters.
- **Y** Change fuel filters.
- **Y** Service and / or replace air filter element.
- **Y** Perform engine oil analysis.
- **Y** Engine tune-up.

Generic Cooling Tower

A (During start-up) Clean and brush down tower, louvers and basin
A (During start-up) Check condition of paint and repair as required
A (During start-up) Assure metering orifices and clean and open
A (During start-up) Clean suction system
A (During start-up) Check tower piping for leaks
A (During start-up) Clean all strainers in piping system
A (During start-up) Check condition of electrical wiring, connections and boxes
Q Lubricate motor bearings
A Clean fan
Q Assure fan blade clamps are tight
A Check operation of float valve
W Check fill valve, float, and linkage for proper operation
W Assure suction screens are clear of sludge
Q Check for scale and algae
S (During shutdown) Open fan and pump motor current breakers
S (During shutdown) Shut-off and drain water supply piping, leave drains open, assure no leaks

Automatic Mixing Box, VAV

Special Instructions 1. Review manufacturer's instructions.

Tools & Materials:
1. Tool Group B
2. Control drawings
3. Calibration tools
4. Cleaning equipment and materials. Consult the MSDS for hazardous ingredients and proper PPE.
5. Duct tape
6. Lubricants: consult the MSDS for hazardous ingredients and PPE.
7. Safety goggles

TASKS

1. Check to see that the operating control thermostat and static pressure sensors activate the damper per design specifications. If not, recalibrate. Replace if items are defective with the same type action (direct or reverse action) and range.
2. Clean inside of box.
3. Check constant volume damper for loose or broken parts and clean. See that adjustment has not come loose. Lightly oil moving parts.
4. Check volume regulator motor for freedom of movement and proper operation.
5. Check air duct and connections for air leaks.

Generic Air Dryer, Refrigerated or Regenerative Desiccant

Special Instructions:

1. Schedule this maintenance in conjunction with the maintenance on the associated air compressor.
2. Review manufacturer's instructions.
3. Review the Standard Operating Procedure for "Controlling Hazardous Energy Sources"
4. De-energize, lock out electrical circuits.
5. Comply with the latest provisions of the Clean Air Act And EPA regulations.
6. No intentional venting of refrigerants is permitted. During the servicing, maintenance, and repair of refrigeration equipment, the refrigerant must be recovered.
7. Whenever refrigerant is added or removed from equipment, record the quantities on the appropriate forms.
8. Recover, recycle, or reclaim the refrigerant as appropriate.
9. If disposal of the equipment item is required, follow regulations concerning removal of refrigerants and disposal of the item.
10. If materials containing refrigerants are discarded, comply with EPA regula tions as applicable.
11. Refrigerant oils to be removed for disposal must be analyzed for hazardous waste and accordingly.
12. For refrigerant type units, closely follow all safety procedures described in the MSDS for the refrigerant and all labels on refrigerant containers.

Tools & Materials:

1. Tool group A
2. Stet
3. Filter cartridges
4. Gasket and packing material.
5. Fin comb
6. Self-sealing quick disconnect refrigerant hose fitting
7. Refrigerant recovery/recycle unit
8. EPA/DOT approved refrigerant storage tanks.

TASKS

1. Lubricate valves and replace packing, if necessary.
2. Check dryer operating cycle.
3. Inspect and clean heat exchanger.
4. Check outlet dew point.
5. Clean and lubricate blower.
6. Check automatic blow down devices.
7. Inspect and replace or reinstall inlet features.
8. Refrigerated type
   a. Check traps.
   b. Check refrigerant level and moisture content. If low level or moisture is indicated, check for refrigerant leaks using a halogen leak detector or similar device. If leaks cannot be stopped or corrected, report leak status to supervisor.
   c. Clean and lubricate.
9. Desiccant type
   a. Replace filter cartridges, both pre-filter and after-filter.
   b. Check the inlet flow pressure, temperature, and purge rate.
   c. Check the desiccant and replace if necessary.
   d. Inspect and clean solenoids purge valves, and strainers.

Automatic Transfer Switch

This applies to those devices utilized to automatically switch and electrical power supply from it's normal source to an alternate or emergency power generators, but they can also be used to transfer from one commercial sources to another. Multiple devices may be used where Uninterruptible Power Supply (UPS) systems are installed.

Special Instructions:

1. Review manufacturer's instructions on operation and maintenance.
2. Review the switching diagram and the affected electrical systems diagrams.
3. Verify locations of generator, transfer switches, critical load, and affected operations.
4. Schedule outage with operating personnel and occupant agencies.
5. All tests shall conform to the appropriate ASTM test procedure and the value used as standards shall conform to the manufacturers and ANSI standards specifications.

Tools & Materials

1. Tool group B
2. Micro-ohmmeter
3. Variable AC voltage source (test cable)
4. AC and DC voltmeter
5. Cleaning equipment and materials

TASKS

Checkpoints:

1. Check with affected occupant agencies and request that agency determine what equipment will be de-energized.
2. Turn off automatic transfer switch and generator automatic controls. Tag control switches.
3. Open and tag supply breaker.
4. Open doors on automatic transfer switch and check phase-to-phase and phase-to-ground for presence of voltage.
5. Clean inside of switch cubicle.
6. Tighten all connections, checking for signs of overheating wires.
7. Disconnect wires attached to each phase of the normal supply, that supplies power to the (E-21) under voltage relays. Test the under voltage relays. After testing relays, reconnect wires.
8. Lubricate mechanism bearings, if required
9. Locate and disconnect operating mechanism control wires and, using a remote source of voltage, operate the mechanism.
10. With the mechanism electrically held, use a micro-ohmmeter to check the contact resistance. Make sure the micro-ohmmeter is connected from the normal supply cable connection to the critical load cable connection. Perform the same test on the emergency source.
11. Reconnect the operating mechanism control wires.
12. Clean indicating lenses and change lamps as needed.
13. Restore the transfer switch to normal position.
14. Check with affected occupant agencies for generator operations.
15. Remove tags and energize normal supply breaker, picking up the critical load.
16. Remove tags and place generator controls in the automatic position.
17. Open normal power breaker; the generator should start and the transfer switch should transfer the critical load.
18. Close the normal power breaker; the transfer switch should transfer the load and the generator should shut down after a cool down period.
19. Check with the affected occupant agencies to see that normal services have been restored to all areas.

Bolted Pressure Contact Switch

Special Instructions:

1. Schedule outage with building tenants and all other interested parties
2. Review manufacturer's instructions
3. Schedule PM at same time as PM of Ground Fault Relay
4. De-energize, lock out, and tag circuit.
5. All test shall conform to the appropriate ASTM test procedure and the values used shall conform to the manufacturer's and ANSI Standard Specifications.

TASKS

1. Inspect for physical damage, proper insulation, anchoring, and grounding.
2. Vacuum and clean interior of unit.
3. Clean insulation, arc chutes and inter-phase barriers.
4. Check fuse linkage and element for proper holder and current rating. Record fuse data.
5. Check contact alignment, wipe and pressure. Make necessary adjustments.
6. Perform contact resistance test across each switch-blade and fuse link.
7. Perform insulation resistance test phase to phase and each phase to ground.
8. Record all test and inspection results

Gas Burner

Special Instructions: 1. Review manufacturer's instructions.

Tools and Materials: 1. Tool group C
2. Flue gas analyzer.
3. Clean wiping cloths.

TASKS

1. Check boiler room for adequate ventilation in accordance with AGA burner requirements.
2. Check operation of all gas controls and valves including: manual gas shutoff; petal gas regulator; petal solenoid valve; safety shutoff valve; automatic gas valve; butterfly gas valve, motor, and linkage to air louver; safety petal solenoid.
3. Check flue connections for tight joints and minimum resistance to air flow.
4. Draft regulators should give slightly negative pressure in the combustion chamber at maximum input.
5. On forced draft burners, gas manifold pressure requirements should correspond with modulating valve in full open position and stable at all other firing rates.
6. Take flue gas readings to determine the boiler efficiency. Use the manufacturer's instructions if available. If they are not, obtain a copy before doing this PM. If efficiency is low, check baffling and passes for short circuiting, and boiler for air infiltration. Adjust dampers and controls to optimize efficiency. Tests should be run at the following load points. 100%, 70%, and 40% of rated full load for boilers having metering controls of modulation capacity at these load points.
   b. At the high and low fire rates on boilers equipped with OFF/LOW FIRE/HIGH FIRE control.
   c. At the single firing load points on boilers equipped with OFF/ON controls only.
7. Check burner for flashback and tight shutoff of fuel.
8. Check that operation and adjustments conform to manufacturer's instructions.

Central Control Panel, HVAC

Special Instructions:

1. Schedule maintenance with operating personnel.
2. Obtain and review manufacturer's information for servicing, testing and operating.
3. Obtain "AS BUILT" diagrams of installation

Tools & Materials:

1. Tool Group B
2. Obtain and understand how to use the manufacturer's testing instruments.
3. Cleaning equipment and materials. Consult the MSDS for hazardous ingredients and proper PPE.
4. Lubricants as specified by equipment manufacturer. Consult the MSDS for hazardous ingredients and proper PPE.
5. Lint free cleaning cloths.

TASKS

1. Clean, lubricate and adjust all electro-mechanical components (printers, relays, graphic projectors, command buttons and switches).
2. Test data transmission to and from remote panels and input/output devices. Recalibrate and/or repair.
3. Verify command functions by observing resultant action (on-off, open-close, etc.).
4. Test alarm report devices and subsystems and analyze visual, audible and printed annunciation. Clean, recalibrate, repair or replace defective components.
5. Test scanning system. Repair if necessary. Note: systems incorporating open type relays should be cleaned.
6. Check operating data. Analyze for accuracy

.

Control Panel-Central Refrigeration Unit

Special Instructions:

1. Schedule shutdown with operating personnel.
2. Obtain and review manufacturer's information for servicing, testing, and operating.
3. Obtain "As Built" diagrams of installation.

Tools aand Materials:

1. Tool group B
2. Cleaning materials and equipment.
3. Pressure gauge
4. Temperature analyzer
5. Multi-meter.

TASKS

1. Clean and calibrate all controlling instruments.
2. Clean or replace orifices and contacts.
3. Check for pneumatic leaks and loose wiring and repair.
4. Replace charts, add ink, and check calibration of flow meter, temperature recorders, and kilowatt charts.
5. Check for bad indicator lights and gauges and replace as necessary.
6. Test all controllers and set at proper set points.
7. Check operating data and analyze for proper operation.

Central Packaged Chilled Water Unit

Tools & Materials:

1. Tool group A and B
2. Pressure washer
3. Respirator
4. Gloves
5. Refrigerant recovery/recycle equipment
6. Self-sealing quick disconnect refrigerant hose fittings
7. EPA/DOT-approved refrigerant storage tanks
8. Cleaning materials and equipment. Consult the MSDS for hazardous ingredients & proper PPE.
9. Fin comb
10. Paint brushes
11. Safety goggles
12. Approved refrigerant

TASKS

1. Condenser

· Remove debris from air screen and clean underneath unit.
· Pressure wash coil with proper cleaning solution.
· Straighten fin tubes with fin comb.
· Check electrical connections for tightness.
· Check mounting for tightness.
· Check for corrosion. Clean and treat with inhibitor as needed.
· Clean fan blades.
· Inspect pulleys, belts, couplings, etc.; adjust tension and tighten mountings as necessary. Change badly worn belts. Multi-belt drives should be replaced with matched sets.
· Perform required lubrication and remove old or excess lubricant.

2. Compressor(s)

· Lubricate drive coupling.
· Lubricate motor bearings (non-hermetic).
· Check and correct alignment of drive couplings.
· Inspect cooler and condenser tubes for scale. Clean if required. Leak test tubes using a halogen leak detector or suitable substitute.
· Add refrigerant per manufacturer's instructions if needed.
· Check compressor oil level.
· Run machine; check action of controls, relays, switches, etc. to see that:
  Compressor(s) run at proper settings.
  Suction and discharge pressures are proper.
  Outlet water temperature is set properly.
· Check and adjust vibration eliminators.

3. Controls

· Check operation of all relays, pilot valves, and pressure regulators.
· Check resulting actions of pressure sensing primary control elements such as diaphragms, bellows, inverted bells, and similar devices when activated by air, water, or similar pressure.

4. Motor

· Check ventilation ports for soil accumulations; clean if necessary.
· Clean exterior of motor surfaces of soil accumulation.
· Lubricate bearings according to manufacturer's recommendations.
  a) Remove filer and drain plugs (use zerk fittings if installed).
  b) Free drain hole of any hard grease (use piece of wire if necessary).
  c) Add grease. Use good grade lithium base grease unless otherwise specified.
· Check motor windings for accumulation of soil. Blow out with low-pressure air or vacuum as needed.
· Check hold-down bolts and grounding straps for tightness.
· Remove tags, start unit, and check for vibration or noise.

Cooling Tower Maintenance

Special Instructions:

1. Schedule performance of this PM activity prior to seasonal start-up. Consider the time needed to effect any required repairs.
2. Review the Standard Operating Procedure for "Controlling Hazardous Energy Sources."
3. Review manufacturer's instructions.
4. De-energize, lock out, and tag electrical circuits.
5. Review the Standard Operating Procedure for "Selection, Care, and Use of Respiratory Protection."
6. Properly dispose of any debris, excess oil, and grease.
7. Check the building's asbestos management plans to see if the wet deck panels have been tested for asbestos. If they are suspect but have not been tested, have them tested. Manage asbestos in accordance with the plan.

Tools and Materials:

1. Tool group C
2. Protective coating, brushes, solvent, etc
3. Manufacturer approved lubricants.
4. Respirator
5. Work gloves
6. Safety goggles
7. Cleaning tools and materials.
8. Amp probe and voltmeter.
9. High pressure washer.
10. OSHA approved ladders of appropriate size or scaffolding. Check ladder for defects. Do not use defective ladders.

TASKS

Exterior Structural

1. Inspect louvers for correct position and alignment, missing or defective items, and supports.
2. Inspect casings and attaching hardware for leaks or defects. Check the integrity and secure attachment of the corner rolls.
3. Inspect for loose or rotten boards on wood casings. Examine from the interior. Extensive dam age may require replacement with fiberglass sheeting.
4. Inspect condition of access doors and hinges. Repair as necessary.
5. Inspect the distribution system including flange connectors and gaskets, caulking of headers on counter flow towers, deterioration in distribution basins, splashguards, and associated piping on cross flow towers. If configured with water troughs check boards for warpage, splitting, and gaps.
6. Examine the drain boards for damage and proper drainage. Check the fasteners also.
7. Inspect stairways including handrails, knee rails, stringers, structure, and fasteners for rot, corrosion, security and acid attack.
8. Shake ladder to verify security, and check all rungs.
9. Check the security, rot, and corrosion on walkway treads. Check treads, walkways, and plat forms for loose, broken, or missing parts. Tighten or replace as necessary.
10. Ladders must be checked for corrosion, rot, etc. Verify compliance with Occupational Safety and Health regulations regarding height requirements. Check ladder security.
11. Check fan decks and supports for decay, missing and broken parts, and gaps. Check the security.
12. Fan cylinders must be securely anchored. Check fastening devices. Note any damaged, missing, or corroded items. Watch for wood rot and corrosion of steel. Verify proper tip

clearance between the fan blade and interior of cylinder. Verify compliance with OSHA requirements regarding height. Check its condition.

13. Apply protective coatings as needed on exterior surfaces. Be sure rust and dirt has been removed first.

Interior Structural:

14. Inspect the distribution system piping for decay, rust, or acid attack. Check the condi tion and tightness of connections and branch arms. Observe spray pattern of nozzles if possible and note missing and defective nozzles. Note condition of the redistribution system under the hot water system.
15. Inspect mechanical equipment supports and fasteners for corrosion. Wood structural members in contact with steel should be checked for evidence of weakness. Check for condition of springs or rubber vibration absorption pads, including adjusting bolts, ferrous members, and rubber pads.
16. Check valves and operating condition of fire detection system. Check for corrosion of pipes and connectors. Check wiring of any thermocouple installed.
17. Check drift eliminators and supports. Remove any clogging debris. Replace missing blades.
18. Inspect tower fill for damage, ice breakage, deterioration, and misplaced, missing, or defective splash bars.
19. Examine interior structural supports. Test columns, girts, and diagonal wood members for soundness by striking with a hammer. A high-pitched, sharp sound indicates good wood, whereas a dull sound indicates soft wood. Probe rotted areas with a screwdriver to determine extent of rot. Look for iron rot of metal fasteners in contact with wood. Check condition of steel internals. Check condition and tightness of bolts.
20. Inspect the nuts and bolts in partitions for tightness and corrosion. Look for loose or deteriorated partition boards. Note if partitions are installed so as to prevent wind milling of idle fans. Make sure wind walls parallel to intake louvers are in position. Boards of transit members should be securely fastened. Check condition of wood or steel supports for rot and corrosion.
21. Check wooden cold-water basins for deterioration, warps, splits, open joints, and sound of wood. Inspect steel basins for corrosion and general condition. Inspect concrete basins for cracks, breaking joints, and acid attack.
22. Check all sumps for debris, condition of screens, antiturbular plates, and freely operating drain valves.

Mechanical

1. Check alignment of gear, motor and fan.
2. Inspect fans and air inlet screens and remove any dirt or debris.
    a. Check hubs and hub covers for corrosion, and condition of attaching hardware.
    b. Inspect blade clamping arrangement for tightness and corrosion.
3. Gear box.
    a. Clean out any sludge.
    b. Change oil. Be sure gear box is full to avoid condensation.
    c. Rotate input shaft manually back and forth to check for backlash.
    d. Attempt to move the shaft radially to check for wear on the input pinion shaft bearing.

e. Look for excessive play of the fan shaft bearings by applying a force up and down on the tip of a fan blade.

Note: Some output shafts have a running clearance built into them.

4. Power transmission.
   a. Check that the drive shaft and coupling guards are installed and that there are no signs of rubbing. Inspect the keys and setscrews on the drive shaft, and check the connecting hardware for tightness. Tighten or install as required.
   b. Look for corrosion, wear, or missing elements on the drive shaft couplings.
   c. Examine the exterior of the drive shaft for corrosion, and check the interior by tap ping and listening for dead spots.
   d. Observe flexible connectors at both ends of the shaft.
   e. Inspect bearings, belts, and pulleys for excessive noise, wear or cracking, alignment, vibration, looseness, surface glazing, tension. Replace or repair as required.
5. Check water distribution. Adjust water level and flush out troughs if necessary. Check all piping, connections, and brackets for looseness. Tighten loose connections and mounting brackets. Replace bolts and braces as required.
6. Check nozzles for clogging and proper distribution.
7. Inspect keys and keyways in motor and drive shaft.

Electrical

1. Check the electric motor for excessive heat and vibration. Lubricate all motor bearings as applicable. Remove excess lubricant.
2. Inspect fused disconnect switches, wiring, conduit, and electrical controls for loose connections, charred or broken insulation, or other defects. Tighten, repair or replace as required.
3. Remove dust from air intakes, and check air passages and fans.
4. If there is a drain moisture plug installed, see if it is operational.
5. Check amps and volts at operating loads, recommend pitching of fans blades to compensate.
6. Look for corrosion and security of mounting bolts and attachments.

Cooling Tower, Cleaning

Special Instructions:

1. Perform work before seasonal start-up, before seasonal shutdown, and quarterly during the cooling season.
2. Review the Standard Operating Procedure for "Controlling Hazardous Energy Sources."
3. Review manufacturer's instructions.
4. De-energize, tag, and lock electrical circuits.
5. Review the Standard Operating Procedure for "Selection, Care, and Use of Respiratory Protection."
6. Ensure that there are safe and sturdy ladders and platforms to perform the lifting and cleaning required.
7. If biological growth is excessive, have a qualified water treatment specialist review your treatment program.
8. Check the building 's asbestos management plans to see if the wet deck panels have been tested for asbestos. If they are suspect but have not been tested, have them tested. Manage asbestos in accordance with the plan.

Tools and Materials:

1. Tool group C
2. Pressure washer with hose and nozzle
3. Cleaning tools and materials.
4. Appropriate chemicals and detergents
5. Respirator with acid/gas/mist/HEPA filters.
6. Safety goggles
7. Waterproof clothing
8. Gloves
9. Rubber boots if wet
10. Litmus paper or pH meter
11. Swimming pool test kit.

TASKS

1. Close the building air intake vents within the vicinity of the cooling tower until the clean ing procedure is complete.
2. Shut down, drain, and flush the cooling tower with water. Isolate the cooling tower from the rest of the condenser water system where applicable.
3. Clean the wet deck, remove all debris, and dispose of properly. If the wet deck panels contain asbestos, follow the asbestos management plan for isolation, notification, work practice, and waste disposal.
4. Inspect the tower, the tower basin and holding tank for sediment and sludge, and any biological growth.
5. Using a low pressure water hose or brushes, clean the tower, floor, sump, fill, spray pans and nozzles and removable components such as access hatches, ball float, and other fittings until all surfaces are clean and free of loose material. Porous surfaces such as wooden and ceramic tile towers will require additional cleaning and brushing. Clean cracks and crevices where buildup is not reached by water treatment.
6. Clean all systems strainers and strainer housings.
7. Remove drift eliminators and clean thoroughly using a hose, steam, or chemical cleanser.
8. Check fan and air inlet screens and remove any dirt or debris.
9. Reassemble components, and fill tower and cooling system with water.
10. Monitor the water pH and maintain pH within a range of 7.5 to 8.0. The pH can be monitored with litmus paper or a pH meter.

If a more thorough disinfectant cleaning is needed:

11. Add a silicate-based low or non-foaming detergent as a dispersant at a dosage of 10-25 pounds per thousand gallons of water in the system.
    a. Use a silicate-based low or non-foaming detergent such as Cascade, Calgonite, or equivalent product.
    b. If the total volume of water in the system is not known, it can be estimated to be 10 times the re-circulating rate or 30 gallons per ton of refrigeration capacity.
    c. The dispersant is best added by first dissolving it in water and adding the solution to a turbulent zone in the water system, such as the cooling tower basin near the pump suction.
    d. Contact a professional water treatment specialist for a dispersant that may be safely used without interfering with operation of the system.
12. Add chlorine disinfectant to achieve 25 parts per million of free residual chlorine.
    a. Maintain 10 ppm of free residual chlorine in water returning to the cooling tower for 24 hours.
    b. A swimming pool test kit may be used to monitor the chlorine. Follow the manufacturer's instructions. Test papers such as those used to monitor restaurant sanitizing tanks may also be used.
    c. Monitor every 15 minutes for two hours to maintain the 10-ppm level. Add chlorine as needed to maintain this level.
    d. Two hours after the slug dose or after three measurements are stable at 10 ppm of free residual chlorine, monitor at two-hour intervals to maintain the 10-ppm of free residual chlorine.
    e. Some kits cannot measure 10 ppm. If so, dilute the test sample with distilled water to bring it within the test set range.
13. After 24 hours, drain the system.
14. Adjust bleed, float, and central valve for desired water level.
15. Open any building air vents that were closed prior to the cleaning of the cooling tower.
16. Implement an effective routine treatment program for microbial control.
17. Document all maintenance and cleaning procedures by date and time. Record the brand name and the volume or weight of chemicals used.

Disconnect, Low Voltage

Special Instructions:

1. Schedule outage with operating personnel.
2. De-energize, lock out, and tag electrical circuit.
3. Obtain and review manufacturer's operation and maintenance instructions.
4. All tests shall conform to the appropriate manufacturer's test procedures and the values used as standards shall conform to GSA, and ANSI, specifications.

Tools and Materials:
1. Tool group B
2. Torque wrench
3. Cleaning equipment and material
4. Vacuum
5. Micro-ohmmeter
6. Appropriate lubricants.

TASKS

1. Inspect for signs of overheating and loose or broken hardware.
2. Inspect connections to bus and cables.
3. Clean main contacts, adjust and put a thin film of conductive lubricant on them if recommended by the manufacturer.
4. If the contacts are burned or the switch has overheated, a contact resistance test should be conducted. Adjust the contacts with the highest readings to correspond to the lowest reading contact. A maximum value can be obtained from the manufacturer's instructions.
5. Check the tubes and renewable elements for corrosion, dirt, and tracking. Clean or replace as necessary.
6. Clean entire cubicle with vacuum.
7. Remove tags and return circuit to service.

Doors, Automatic, Hydraulic/Electric or Pneumatic

Tools and Materials
1. Tool Group B
2. Lubricants as specified by equipment manufacturer. Consult the MSDS for hazardous ingredients and proper PPE.

TASKS

1. Check alignment of doors and mechanisms. Inspect mountings, hinges, mats, and trim, weather stripping, etc. Replace, tighten, and adjust as required.
2. Operate with power, observing operation of actuating and safety mats, door speed, and checking functions.
3. Check manual operation.
4. Inspect power unit, lubricate and tighten lines as required.
5. Check operation of control board relays, clean, replace, adjust contacts as required.
6. Inspect door operating unit, tighten lines, and adjust as required.
7. Clean and lubricate door pivot points.
8. On pneumatic or hydraulically operated door operators, check for correct operating pressures per manufacturer’s instructions.
9. Clean up and remove all debris from work area.

Doors, Power Operated

Special Instructions:

1. Review manufacturer's instructions.

Tools and Materials:

1. Standard Tools – Basic
2. Cleaning equipment and materials. Consult the MSDS for hazardous ingredients and proper PPE.
3. Lubricants as specified by equipment manufacturer. Consult the MSDS for hazardous ingredients and proper PPE.

TASKS

1. Inspect general arrangement of doors and mechanisms, mountings, guides, wind locks, anchor bolts, counterbalances, weather stripping, etc. Clean, tighten, and adjust as required.
2. Operate with power from stop to stop and at intermediate positions. Observe performance of various components, such as brake, limit switches, motor, gearbox, etc. Clean and adjust as needed.
3. Check operation of electric eye, treadle, or other operating devices. Clean and make required adjustments.
4. Check manual operation. Note brake release, motor disengagement, functioning or hand pulls, chains sprockets, clutch, etc.
5. Examine motor, starter, push button, etc., blow out or vacuum if needed.
6. Inspect gearboxes, change or add oil as required.
7. Perform required lubrication. Remove old or excess lubricant.
8. Clean unit and mechanism thoroughly. Touch up paint where required. Clean and remove all debris.

Elevator, Electric or Hydraulic - Semiannual

Tools and Materials:

1. Standard Tools Basic
2. Cleaning tools and materials
3. Out of service signs
4. Barricades
5. Lubricants

TASKS

1. Cables: Inspect, lubricate, and properly adjust hoist cables, compensating cables, governor cables, and traveling cables to their manufacturer's specifications. Check all cable fastenings. Inspect guide rails and counterweight. Check and adjust the slow down and limit switches. Adjust all other items as necessary to obtain proper equipment operation.
2. Sheaves: Inspect, clean, and lubricate in accordance with manufacturer's specifications all deflector, compensating, and top of car sheaves.
3. Motors: Inspect connections, armature and rotor clearances on hoist motor and motor generator set; than clean and adjust as necessary to obtain proper operation.

Elevator, Electric or Hydraulic

Special Instructions: Check manufacturer's instructions, those that have more stringent guide lines for preventive maintenance shall be followed. The frequencies shown here are minimum requirements and are in addition to the regular PBS inspection tour. Items regularly inspected on a weekly basis include the motor-generator unit, hoist machine, controls, and governor. Doors, hangers, closers, interlocks, door operators should be checked frequently for proper operations by qualified elevator mechanics or inspectors as they ride the elevators. Items requiring attention should be reported to the elevator shop supervisor or elevator contractor. This guide includes checkpoints that should be accomplished on an annual basis as noted.

Tools and Materials:

1. Standard Tools-Basic
2. Cleaning tools and materials
3. Out of Service signs
4. Barricades
5. Lubricants

TASKS

1. Brakes: Completely dismantle brake assembly, clean and inspect for wear. Replace defective parts to obtain proper inspection. Lubricate bearing, pins, and pivot points.
2. Selector: Inspect, clean, lubricate, replace parts, and make repairs or adjustments as necessary for proper operation of selector unit including cables, chains, clutches, cams, gears, fuses, motor brushes, wiring, connections, contacts, relays, tapes, tape tension, sheave, broken tape switch, and tape wipers.
3. Controller: Thoroughly clean controller with blower or vacuum. Inspect and check operation of switches, relays, timers, capacitors, resistors, contacts, overloads, wiring, connections, fuses, overload oil levels, and overload control. Check for MG shutdown, high call reversal, zone control, and load by-pass door failure time. Check programming up peak, down peak, off hours, and off peak. Replace defective parts and adjust controller for proper operation.
4. Hoist way Doors: Clean, inspect and lubricate all door operating mechanisms, including but not limited to rollers, up thrusts, interlocks, clutches, self-closing gibs, and sills. Replace worn parts, repair or adjust door mechanisms as necessary to obtain proper operation.
5. Hoist ways: Clean rails, beams, and all related ironwork in hoist way. Dust hoist way walls. Clean top, bottom, and sides of car. Clean counterweight.
6. Hoist Machine and Motor-Generator: Clean with blower or vacuum. Clean end bells, brush riggings, and commutator.
7. Buffers: Check oil level and operations of switches. Add oil or adjust switches as necessary for proper operation. Manually compress buffer and test for proper return in accordance with ASME/ANSI A17.1 Safety Code for Elevators and Escalators, Rule 201.4e(1).
8. Scheduling, Dispatch and Signal Boards: Clean with blower or vacuum. Inspect and check operation of switches, relays, timers, capacitors, resistors, contacts, overloads, wiring, and connections. Replace worn parts and adjust controller for proper operation.
9. Motors: Change oil in hoist motor, MG set, geared machines, and gear boxes with lubricants as specified by the equipment manufacturers(s).
10. Safeties: Inspect, clean, lubricate, and manually operate safety mechanisms prior to slow speed safety test. Replace parts or adjust as necessary to obtain proper operation of safety devices.

Fan, Centrifugal

Special Instructions:
1. Review manufacturer's instructions.
2. Schedule shutdowns with operating personnel, as needed.
3. De-energize, lock out, and tag electrical circuits.
4. If the fan motor is 1hp or larger, schedule PM on the motor at same time.

Tools and Materials:
1. Standard tools - basic
2. Tachometer
3. Cleaning equipment and materials.
4. Vacuum
5. Grease guns, lubricants.
6. Respirator.

TASKS
1. Check fan blades for dust buildup and clean if necessary.
2. Check fan blades and moving parts for excessive wear. Clean as needed.
3. Check fan RPM to design specifications.
4. Check bearing collar set screw on fan shaft to make sure they are tight.
5. Vacuum interior of unit if accessible. Clean exterior.
6. Lubricate fan shaft bearings while unit is running. Add grease slowly until slight bleeding is noted from the seals. Do not over lubricate. Remove old or excess lubricant.
7. Check belts for wear, adjust tension or alignments, and replace belts when necessary. Multiple belt drives should be replaced with matched sets.
8. Check structural members, vibration eliminators and flexible connections.
9. Remove all trash and clean area around fan.

Filter

Special Instructions:
1. De-energize, lock out, and tag fan electrical circuit.
2. Filters should be replaced when static pressure reading indicates or by schedule

Tools and Materials
1. Standard Tools - Basic
2. Respirator
3. Vacuum
4. Filter replacement

TASKS

Throw-away
1. Remove old filters.
2. Vacuum filter section of air handler.
3. Inspect frame, clamps, etc.
4. Install new filters. Make sure direction of airflow corresponds to the airflow shown on the filter and filters are properly sized to cover the opening.
5. Remove tags, and restore to service.
6. Clean up work area and remove trash.

Viscous Type (wire mesh)
1. Remove filters and replace with filters that have been cleaned and recoated. Examine frame and clean it with a high suction vacuum.
2. Move dirty filters to cleaning station.
3. Clean, recoat, and store filters removed until next scheduled change.

Fire Alarm Box (Manual-Coded and Un-coded)

Special Instructions:
The work required by this procedure may cause the activation of an alarm and/or supervisory signal. The field office manager and the control center or fire department that will receive the alarm and/or signal must be notified prior to start of work. When alarm systems are connected to municipal systems, test signals to be transmitted to them will be limited to those acceptable to that authority. Results should be recorded.

TASKS

1. Examine box for damage and legible box number.
2. Check external tamper devices.
3. When practical, remove "Break Glass" or glass rods and follow instructions for actuating alarm.
4. Confirm that proper signal is transmitted to receiving station.
5. Determine that audible alarms or signals, local or general, and actuated by the alarm box are operating.
6. General- Check other features for activation by stations or boxes through the fire alarm control panel. These features include alarm bells, elevator capture, releasing of fire doors held open, notification of fire department, smoke control, etc.
7. Inspect recording register for legibility, time, code number, and number of rounds.
8. On systems with shunt non-interfacing or positive non-interfacing circuits, operate one box and then another box on each box loop prior to the completion of the first cycle. Check for interference at receiving station or recording register.
9. Restore alarm box and accessories to normal position promptly after each test. This restoration may include rewinding, resetting, replacement of tamper devices, etc.

Fire Department Hose Connection

Standard Instructions:
The work required by this procedure may cause the activation of an alarm and/or supervisory signal. The field office manager and the control center or fire department that will receive the alarm and/or signal must be notified prior to start of work. When cracking valve, do not stand directly in front of opening.

TASKS

1. Remove obstructions to easy accessibility of hose connection.
2. Inspect cut off valves and check valves (usually located at base of standpipe riser) for corrosion or leakage. Exercise cut off valve and repack if necessary.
3. Remove cap from hose connection and check threads.
4. Crack valve until water sweeps through valve. Then close valve and check for leaks.
5. Screw cap onto valve until it is hand-tight.
6. Clean up work area and remove all trash.

Fire Door, Stairwell and Exit way (swinging)

Standard Instructions:
The work required by this procedure may cause the activation of an alarm and/or supervisory signal. The field office manager and the control center or fire department that will receive the alarm and/or signal must be notified prior to start of work.

TASKS
1. Remove all hold-open devices such as fusible links except approved electro-magnetic hold open devices.
2. Check hang and swing for close fit. Doors must latch on normal closing cycle and have a clean neat fit.
3. Remove any obstructions that retard full swing or movement of door.
4. Test operation of panic hardware.
5. Inspect door coordinates on pairs.
6. Check operation of any special devices such as smoke detectors or magnetic door releases.
7. Inspect doors for damage.
8. Clean up work area and remove all trash.

Fire Extinguisher Hydrostatic Testing, CO2, Store

Special Instructions:
Soda acid, carbon dioxide, and foam extinguishers should be tested on a 5-year basis. Dry chemical extinguishers, with the exception of those with stainless steel shells should be tested on a 12-year basis. Testing should be in accordance with NFPA Standard No. 10. Hydrostatic testing of extinguishers requires experienced personnel and suitable testing equipment.

Tools and Materials:
1. Standard tools - basic
2. Seals
3. Appropriate testing equipment.
4. Tags.
5. Scale

TASKS
1. Any cylinders that have been repaired by soldering or welding, damaged, corroded, burned, or had calcium chloride type of extinguishing agent used in stainless steel extin guisher shall not be hydrostatically tested, but destroyed.
2. Operate stored pressure and cartridge type extinguishers and check performance.
3. Dismantle and remove all traces of extinguishing agent from inside of shell and hose assembly.
4. Insert plug into shell opening.
5. Fill with water and connect the test pump
6. Secure shell in protective cage and apply proper test pressure. Pressure to be applied at rate so test pressure within one minute.
7. Observe shell and gauge for any distortion or leakage.
8. All dry chemical and dry powder extinguishers must have all traces of water removed from extinguishing agent, shell, hose, and nozzle. A heated air stream is recommended with its temperature not exceeding 150 degrees F.
9. Weigh replacement cartridge to insure that it is full of gas.
10. Recharge extinguisher according to manufacturer's instructions.
11. Affix permanent record on extinguisher with note of year of hydrostatic test.

Fire Pump, Motor or Engine Driven - Annual

Standard Instructions
The work required by this procedure may cause the activation of an alarm and/or supervisory signal. The office and the control center or fire department that will receive the alarm and/or signal must be notified prior to start of work. It is recommended that a yearly test shall be made at full pump capacity and over to make sure that neither pump nor suction pipe is obstructed.

Tools and Materials:

1. Tool group C
2. Coolant. Consult the MSDS for hazardous ingredients and proper PPE.
3. Engine oil. Consult the MSDS for hazardous ingredients and proper PPE.
4. Oil, air, fuel filters.
5. Cleaning equipment and materials. Consult the MSDS for hazardous ingredients and proper PPE.
6. Tune-up kit.

TASKS
ANNUAL

1. Engine
   a. Change crankcase oil.
   b. Flush cooling system and check hoses, replace coolant.
   c. Clean air and fuel filters, replace when needed.
   d. Tune engine.
   e. Increase RPM until over-speed or governor operates.
   f. Check for proper operation of speed controller.
   g. Check for alignment and vibration.
   h. Adjust clutch.
2. Perform other work prescribed by manufacturer.
3. Motor Refer to PM guide for motor PM steps.
4. Clean up work area and remove all trash.

MONTHLY

1. Inspect for dirt collected at bleed port and restriction elbow. Clean if necessary.
2. Inspect joints for leakage. Tighten all bolts.
3. Check for dust or other material that may have sifted onto the upper face of the pilot pressure plate.
4. Remove and clean line strainer (back-flush where possible).
5. Inspect valve head and seats for nicks and abrasions. Notify supervisor if valve requires regrinding.
6. Inspect pressure reading against set point.
7. Check for free operation of valve stem
8. Inspect condition of diaphragm.
9. Inspect pilot line for leaks.
10. Clean up work site and remove all debris.

Grease Trap

Special Instructions: Use appropriate protective clothing, especially safety glasses.

Tools and Materials
1. Standard Tools Basic
2. Gloves
3. Goggles

TASKS
1. Clean out trap and sterilize.
2. Inspect for clogging, scale, and improperly positioned or missing baffles.
3. Tighten loose parts as necessary

Humidification System

Special Instructions:
1. Review manufacturer's instructions.
2. Review the Standard Operating Procedure for "Selection, Care and Use of Respiratory Protection"
3. Turn off water supply
4. Secure electrical service before servicing humidification system, if applicable.
5. Use of work gloves may be necessary due to caustic residual mineral deposits.

Tools and Materials:
1. Tool group A
2. Psychrometer
3. Coil cleaning equipment
4. Work gloves
5. Safety goggles
6. Respirator.

TASKS
1. Operate humidistat through its throttling range to verify activation, or deactivation of humidifier.
2. Clean and flush condensate pans, drains, water pans, etc. Remove corrosion, and repaint as needed, ensure that it does not become a part of the indoor air by creating large amounts of volatile organic compounds or irritants. Check the MSDS to see what haz ardous products are present. If hazardous products are present, rinse very well before the system is returned to use. Ensure that the paint lead level is 0.06% or less.
3. Check condition of heating element. Clean steam coils.
4. Clean steam/water spray nozzles. Adjust/replace as needed.
5. Chemically clean exterior of coil to remove scale and encrustations.
6. Inspect steam trap for proper operation.
7. Inspect pneumatic controller for air leaks.
8. Inspect water lines for leaks and corrosion. Tighten all connections and repair leaks.

Heat Exchanger, Flat Plate

Application: This PM guide applies to all flat plate heat exchangers used in central chiller plants for free cooling.

Special Instructions:

1. Review manufacturer's instructions regarding manual cleaning and cleaning-in-place procedures. Where possible, it is recommended that a cleaning-in-place system be uti lized which will allow pumping water or cleaning solution into the unit without disassembly.
2. Obtain operating logs.
3. Review operating logs to check loss of efficiency of heat exchange surfaces, indicating scale and /or corrosion buildup.
4. Never use a steel brush or steel wool on the plates of the heat exchanger. If a brush is needed, use a fiber-type brush.
5. Do not scratch the gasket material during PM.
6. Do not open the unit when hot.
7. Always use clean water (free from salt, sulfur, or high iron concentrations) for flushing and rinsing operation.
8. If steam is to be used as a sterilizing media, do not exceed 270 deg F steam temperature with nitrile gaskets or 300 deg F with EPR gaskets.
9. If chlorinated solutions are used as the cleaning media, they should be at minimum con centration at the lowest temperature possible with the minimum exposure time to the plates. Do not exceed chlorine content levels of 100 ppm at temperatures not higher than 100 deg F with a maximum exposure time to the plates of 10 minutes.
10. Do not inject concentrated cleaning solutions directly into the unit. Add to water first.
11. Do not use hydrochloric (muriatic) acid for cleaning plates.
12. If a cleaning -in-place system is to be used, it must be used before full fouling of the unit can occur.

Tools and Materials:

1. Standard tools-Basic
2. Cleaning solutions. Consult the MSDS for hazardous ingredients and proper PPE.
3. Hose
4. Fiber brush
5. Clean, dry cloth
6. Solution pumping system

TASKS

Manual Cleaning:

1. Open the unit in accordance with the manufacturer's instruction
2. Clean each plate separately. Depending upon the amount of cleaning to be performed, the plate can be cleaned while still hanging or removed and placed on a flat surface to be cleaned.
3. Brush each plate. If rusted or pitted areas appear on the plates, clean up the areas with commercial scouring powder; rinse each plate thoroughly with clean water.
4. Wipe gaskets dry with a clean, dry cloth, removing solid particles, which may cause dam age or leakage.
5. Inspect the lower portion of each plate carefully and clean appropriately. This is the pri mary area where residual solid material will accumulate.

6. Wipe off the mating surface, i.e., the rear of the plate where the gasket seats.
7. Reassemble the unit in accordance with the manufacturer's instructions.
8. Place the unit in service. Inspect thoroughly for leaks.

Cleaning-In-Place:

1. Drain both sides of the unit. If the unit cannot be drained, push systems liquids out of the unit with flush water.
2. Flush unit on both sides with warm water (110 deg F) until the effluent water is clear and free of system fluids.
3. Drain the flush water from the unit and connect a cleaning solution pump assembly to the unit.
4. Flow cleaning unit solutions from bottom to top to insure wetting of all surfaces with solution. If multiple pass unit, reverse flow for at least 1/2 of the cleaning time to wet all surfaces.
5. For optimum cleaning efficiency, the flow rate of water, rinse, and/or cleaning solution should be greater than the normal system fluid flow rate.
6. Flush thoroughly with clean water after completion of the cleaning solution flush.

Lighting, Outside, Incandescent, Fluorescent, etc

Application and Special Instructions:

This guide applies to parking lot, street, loading dock, and perimeter lighting, and provides for group relamping and maintenance of such fixtures outside the building.
Review the Standard Operating Procedure for "Controlling Hazardous Energy Sources."

Tools and materials

Standard Tools-Basic
Ladder. Check ladder for defects. Do not use defective ladders.
Cleaning materials. Consult the MSDS for hazardous ingredients and proper PPE.

TASKS

1. Open and tag switch.
2. Remove old lamp and clean fixture including reflector, refractor, and globes.
3. Inspect condition of wiring, contacts, terminals, and sockets. Look for evidence of overheating.
4. Install new lamp and assemble checking gaskets for proper seat.
5. Test operation of automatic switches.
6. Inspect lamp standards and mounting devices.
7. Clean up work area and remove all trash.

Loading Ramp, Adjustable

Special Instructions:
1. De-energize, lock out, and tag electrical circuit.
2. Review manufacturer's instructions.
3. Review the Standard Operating Procedure for "Controlling Hazardous Energy Sources."

Tools and Materials
1. Tool Group C
2. Hydraulic fluid
3. Lubricants. Consult the MSDS for hazardous ingredients and proper PPE.
4. Cleaning materials. Consult the MSDS for hazardous ingredients and proper PPE.

TASKS
1. Inspect structural features, framework, support members, anchor bolts, pits, platform, etc. Examine condition of bumper. Does it protect ramp properly?
2. Remove dirt and trash from pit and determine if pit drain is open.
3. Inspect motor, controls, starter, push buttons, solenoids, etc. Clean, adjust and lubricate as necessary.
4. For Hydraulic Units:
   · Inspect coupling, pump, control valves, piping, relief valve reservoir, fill pipes, cap, vents, etc. Clean, adjust and lubricate as needed.
   · Inspect cylinder, ram, packing glands, etc. Add or renew packing as required.
   · Change oil as required. Review the MSDS for disposal of used oil. If appropriate, recycle oil at an authorized station. Contact the Regional S and EM office if you have any questions.
5. For Electro-Mechanical Units:
   · Clean and inspect coupling, reduction gear, sprockets and chain, gear trains, screw and lever, and/or other mechanical features. Look for misalignment, loose bolts, evidence of binding or wear, excessive clearance, etc., and tighten as necessary.
   · Examine lubrication devices. Service if required.
   · Test operation of ramp in all directions using a load if possible. Note if ramp holds and does not creep when load is applied or removed. Adjust if necessary.
   · Check manual operation, power disengagement, etc.
   · Lubricate as required.
   · Clean up work area.

Motor Starter, 100 HP and Up

Special Instructions:

1. Schedule outage with operating personnel.
2. Obtain and review manufacturer's instructions for starter to be tested.
3. De-energize, lock out, and tag electrical circuit.
4. All tests shall conform to the appropriate ASTM test procedure and the values used shall conform to the manufacturer's and ANSI Standards specifications

Tools and Materials:

1. Tool group B
2. High Current Test set
3. Micro-Ohmmeter
4. Megger
5. Cleaning equipment
6. Vacuum.

TASKS

1. Visually inspect for broken parts, contact arcing or any evidence of overheating.
2. Check motor nameplate for current rating, and controller manufacturer's recommended heater size.
3. Check line and load connections, and heater mounting screws for tightness.
4. Perform time/current characteristics test at the appropriate multiple of heater rating.
5. Record test results. Show both as found and as left.
6. Check contact resistance in micro-ohms and dielectric strength in meg-ohms
7. Check starter connection by applying a thin film of black contact grease to the line and load stabs, then rack the breaker in and out of the cubicle and measure the wipe marks on the stab. Clean contacts.
8. Remove tags and lock, return circuit to service.
9. Clean work area.

Motor Control Center

Special Instructions:

1. Schedule outages with operating personnel.
2. Review manufacturer's instructions.
3. De-energize, lock out, and tag all electrical circuits.
4. All tests shall conform to the appropriate ASTM test procedure and the values used as standards shall conform to the manufacturer's and ANSI Standards specifications.

Tools and Materials:

1. Tool group B
2. Cleaning equipment and materials
3. Lubricants
4. Vacuum cleaner
5. Contact burnishing tool.

TASKS

1. Tighten all connections to main bus.
2. Inspect breakers and fuses connected to the main bus for tightness.
3. Inspect starter coils. Clean contacts, replace as required.
4. Use vacuum or dry compressed air to remove dust or other material, which may cause shorts or arcing.
5. Inspect all interlocks and controls. Clean and lightly lubricate friction points. Remove excess lubricant.
6. If applicable, inspect contactor/switch arc chutes for cracks or pitting. Repair and clean as needed.
7. Test starter heaters for correct design amperage and size.
8. Test main breaker or fuses to M.C.C. for correct voltage drop, and amperage draw.
9. Operate breakers to insure proper making.
10. Open the starter cover and place the starter in the "test" or "safe" position. Energize the starter.
11. Look for arcing or improper contacting.
12. Visually check coils and contacts. Clean the contacts if needed.
13. Clean the starter interior with dry compressed air.
14. Tighten all connections. Make sure all electrical connections and contacts are properly made between the control apparatus and the motor.
15. Carefully lubricate the friction points on the moving parts of the starter and wipe off excess lubricant.
16. Remove tags and locks, return circuit to service.
17. Clean work area.

Motor Starter, 5hp to less than 100hp and less than 60

Special Instructions:

1. Schedule outage with operating personnel.
2. Obtain and review manufacturer's instructions.
3. De-energize, lock out, and tag electrical circuits.

Tools and Materials:

1. Tool group B
2. Cleaning equipment and materials
3. Vacuum cleaner
4. Electrical contact lubricant
5. Ladder.

TASKS

1. Visually inspect for broken parts, contact arcing, or any evidence of over heating.
2. Check motor nameplate for current rating and controller manufacturer's heater size.
3. Check line and load connections for tightness.
4. Check heater mounting screws for tightness.
5. Check all control wiring connections for tightness.
6. On all units equipped with motor reversing capacity, check mechanical interlocks.
7. On units equipped with two stage starting, check dashpots and timing controls for proper operation. Adjust as required.
8. On units equipped with variable speed starters:
   a. check tightness of connections to resistor bank.
   b. Check resistor coils and plates for cracking, broken wires, mounting and signs of over heating. Clean if required.
   c. Check for tightness of connections to drum controller.
   d. Check contacts of drum controller for arcing and over heating. Apply a thin film of lubricant to drum controller contacts and to rotating surfaces.
9. Check for starter contact connections by applying a thin film of black contact grease to line and load stabs, operate contacts and check surface contact.
10. Lubricate all moving parts with proper lubricant.
11. Clean interior and exterior of cabinet.
12. Energize circuit and check operation of starter and any pilot lights. Replace as required.

Motors, Electric, 1HP or more Special Instructions:

1. If necessary, schedule shutdown with operating personnel.
2. Review manufacturer's Instructions.
3. Review the Standard Operating Procedure for "Controlling Hazardous Energy Sources."
4. De-energize, lockout, and tag electrical circuit serving motor, when applicable. Tools and

Materials:
1. Tool group B
2. Tachometer
3. Cleaning equipment
4. Lubricants
5. Wheatstone bridge
6. Surge tester
7. Megger
8. Voltmeter
9. Capacitance Measurement Bridge
10. Amp meter
11. Power factor meter.

TASKS
1. Check ventilation ports for soil accumulation, clean if necessary.
2. Clean exterior of motor surfaces of soil accumulation.
3. Lubricate bearings according to horsepower ratings:

HP Range Frequency	over 7.5 to 50 HP yearly
1-7.5 HP Every 4 Years	over 50 HP two times/year

4. Check motor windings for accumulation of soil. Blow out with air if required
5. Check hold-down bolts and grounding straps for tightness.
6. Remove tag and lock, energize, and return to service. Clean up work area.

Predictive Maintenance Check Points: The following electrical tests are to be done on motors rated at 10 hp and greater, and are to be accomplished at the motor control panel and should be completely non-destructive.

1. The electrical circuits and the motors shall be non-destructively tested from the load side or the secondary side of the breaker.

Testing shall be done to establish the present operating parameters of the wiring and the motors for the following aspects:

a. Resistance imbalance with results expected to be less than 0.05 ohms in each phase; per NEMA MGI - 14.35 Note: A Wheatstone bridge tester will give these results.
b. Total inductance imbalance with results expected to be less than 35% from a phase-to-phase analysis on the system. Note: a surge tester will give these results.
c. Leaks to ground with results expected to be greater than 5 meg ohms in each phase per IEEE 43-1974, pg. 93. Note: a Megger may be used to give this result.
d. Report on any visual findings of significance or conditions found from testing that need further investigation.
e. Three-phase dynamic testing of AC motors in operation will be done on all systems operating at 600 VAC or less. Record each phase voltage balance with results expected to be less than 1% imbalance per EASA Guide Book, pg. 18. Note: Voltmeters will give these results.
f. Capacitance imbalance when capacitors are part of the installation, with results expected to be less than 10% imbalance. Note: "Capacitance Measurement Bridge" will give these results.
g. Record amps at full load or at maximum design load to be on system with results expected to be less than nameplate full load amps. Note: An ammeter will give this result
h. Record the power factor of the system under load, using a power factor meter.

2. Compare the results of each test performed in step 1 with the previous year's results and consid er how serious the combinations of problems are, and what priority they have for repair or correction.
3. Restore all equipment, as it was when this work was started. Remove tags and return to service. Clean up work area.

Non-destructive Chiller Tube Analysis

Special Instructions and Application:

This PM guide applies to all central refrigeration and central package chilled water units.

1. Coordinate performance of this PM activity with performance of annual PM on the central or package chilled water unit (PM guides R-03, R-04, R-05, R-06, R-07, as applicable).
2. Complete an eddy current test of all heat exchanger tubes, both evaporator and condenser (if applicable), plus concentrator and absorber in absorption units.
3. The test shall be performed in accordance with current requirements and procedures of the American Society of Mechanical Engineer (ASME) Boiler and Pressure Vessel Code Section V, Nondestructive Examination, Article 8, Eddy Current Examination of Tubular Products, and applicable recommended practice standards of the American Society for Testing and Materials (ASTM) for Eddy Current Testing.
4. A Certified Level II or higher technician or equivalent shall conduct this analysis in accordance with the American Society of Non-destructive Testing Recommended Practices, SNT-TC-1A, current version.
5. The test is to be witnessed by the Contracting Officer's Representative or designated inspector.

Reports and Records:

1. A copy of the magnetic tape record shall be maintained by the NDT contractor and furnished if requested by the Government.
2. A preliminary job site report shall be provided as soon as the test is completed.
3. Within ten (10) working days following completion of the test, the NDT contractor shall provide two complete test reports. Include the following:
   a. Written test procedure.
   b. Recommendations-List all tubes recommended for replacement or isolation.
   c. Make complete description of defects (location, depth, inside or outside surface).
   d. Map location- Show tube row, number, and support for each tube bundle.
   e. Name of technician performing tests and evaluating data.
   f. Contractor's certification of technician qualifications TASKS Procedure:

1. Prepare equipment for non-destructive testing (NDT). Remove heat exchange heads, piping, clean tubes, and erect scaffolding as necessary.
2. Test shall be recorded as required by the ASME code Section V (Article 8 – Appendix I, Article I-20).
3. System calibration shall be confirmed hourly.
4. The written procedure in paragraph I-23, Article 8 - Appendix I, in the ASME code is required to be followed.
5. Strip chart recordings shall be provided for:
   a. Each calibration standard and artificial discontinuity comparator used. Annotate to identify each defect machined in the standard and calibration of each division on the chart.
   b. Typical good tube in each bundle.
   c. For each defective tube, annotate to identify tube. Indicate nature and extent of defect.
6. Test each tube to detect, as a minimum, leaks, saddle damage, pitting, interior erosion/corrosion, gasket condition, presence of "tramp" metal, presence of tube bulges, tube seam condition, visual inspection of scale buildup, and tube sheet condition.
7. Correct deficiencies as directed.
8. Restore equipment to service.

Parking Arm Gates

Special Instructions: 1. Obtain and review manufacturer's instructions

Tools & Materials:

1. Tool group B
2. Torque wrenches
3. Cleaning equipment and material. Consult the MSDS for hazardous ingredients and proper PPE.
4. Asphalt filler. Consult the MSDS for hazardous ingredients and proper PPE.
5. Appropriate lubricants. Consult the MSDS for hazardous ingredients and proper PPE.

TASKS

1. Lubricate mechanism with graphite.
2. Adjust linkage between motor and arm.
3. Check and adjust arm pressure.
4. Check and adjust sensitivity on magnetic coils embedded in asphalt.
5. Fill cracks in asphalt where coils are imbedded.
6. Clean and adjust electric breakers.

Pump, Centrifugal

Special Instructions:

1. Review manufacturer's instructions.
2. Pump maintenance should be scheduled to coincide with drive motor maintenance.

Tools and Materials:

1. Tool group C
2. Alignment indicator
3. Grease gun
4. Cleaning materials
5. Hoist assembly for large pumps.

TASKS

1. Check that base bolts are securely fastened.
2. After shutdown, drain pump housing, check suction and discharge valves for holding.
3. Remove cover, gland and packing.
4. Remove corrosion from impeller shaft and housing cover.
5. On pumps with oil ring lubrication, drain oil, flush, and fill to proper oil level with new oil.
6. Inspect wear rings, seals, and impeller.
7. Clean strainers.
8. Replace packing, reassemble
9. Start and stop pump, noting vibration, pressure and action of check valve.
10. Lubricate impeller shaft bearings do not over lubricate.
11. Check drive shaft coupling.
12. Check motor and pump alignment.

    Coupling Size Allowable Alignment
    1"-2" 0.101 total indicator reading
    over 2"-4" 0.015 total indicator reading
    over 4"-7" 0.020 total indicator reading

Refrigeration Controls, Central System

Special Instructions: 1. Read and understand the manufacturer's instructions.
2. Obtain "As Built" diagrams of the control and safety systems.
3. Replace defective control safeties found while performing preventive maintenance.

Tools and Materials:

1. Tool group B
2. Pneumatic Control Gauge
3. Volt Ohm Meter
4. Manufacturer's Control Kit.

TASKS

1. Check flow or pressure differential switches for proper operation. Calibrate/replace as necessary.
2. Check oil temperature control and safety for proper operation. Calibrate/replace as necessary.
3. Check set point of low temperature control and safety for proper operation. Calibrate as necessary.
4. Check capacity controller or demand limiter for proper operation. Calibrate/replace as necessary.
5. Check oil pressure control and safety for proper operation. Calibrate/replace as necessary.
6. Check high-pressure shutout for proper setting and operation. Calibrate/replace as necessary.
7. Check and clean all electrical contacts and pneumatic orifices.
8. Check pneumatic tubing for leaks or damage. Repair or replace as required.
9. Check electrical wiring insulation and connections. Tighten or replace if necessary.
10. Check damper or unloader controller for proper operation. Check position of damper for proper operation. Calibrate/replace as necessary.
11. Check all settings and set points with manufacturer's instructions.

Remote Air Intake Damper

Tools and Materials:

1. Standard Tools - Basic
2. Cleaning equipment and materials. Consult the MSDS for hazardous ingredients and proper PPE.
3. Lubricants: consult the MSDS for hazardous ingredients and PPE. TASKS Checkpoints:
    1. Check damper for freedom of movement and proper operation.
    2. Observe damper operation through full operating range by activating controller. Adjust link age on vanes if out of alignment.
3. Check damper surfaces for wear and clean vanes.
4. Check actuator/damper linkage for proper operation. Adjust if needed. Tighten operator arm set screws.
5. Lubricate mechanical connections sparingly. Wipe off excess.
6. Check actuator for proper operation. If it does not stroke properly, check for binding drive stem. If actuator still does not operate properly replace the diaphragm (pneumatic actuators).
7. Check for air leaks around actuator and in air line between controller and actuator.
8. Lubricate actuator linkage sparingly. Wipe off excess lubricant. DO NOT LUBRICATE actuator/drive stem.
9. Clean off any corrosion or rust on damper frame and/or damper blades, coat with proper type and color paint.

Refrigeration Machine, Centrifugal

Special Instruction:

1. Review manufacturer's instructions.
2. Coordinate PM of refrigeration machine control panel and refrigeration machine controls in conjunction with this activity.
3. Review the Standard Operating Material Procedure for "Controlling Hazardous Energy Resources"
4. De-energize, lockout, and tag electrical circuits.
5. The replacement filter-drier cores for the high efficiency purge unit absorb water vapor from ambient air, so they are shipped in sealed containers. Don't open them until the cores can be installed and sealed in the purge tank.
6. Comply with the latest provisions of the Clean Air Act and Environmental Protection Agency regulations as they apply to protection of stratospheric ozone.
7. No intentional venting of refrigerants is permitted. During the servicing, maintenance, and repair of refrigeration equipment, the refrigerant must be recovered.
8. Whenever refrigerant is added or removed from equipment, record the quantities on the appropriate forms.
9. Recover, recycle, and reclaim the refrigerant as appropriated
10. If disposal of the equipment item is required, follow regulations concerning removal of refrigerants and disposal of the appliance.
11. If materials containing refrigerants are discarded, follow regulations concerning hazardous waste where applicable.
12. Refrigerant oils removed for disposal must be analyzed for hazardous waste and handled accordingly.
13. Closely follow all safety procedures described in the MSDS for the refrigerant and all labels on refrigerant containers.

Tools & Materials:

1. Tools groups A and C
2. Gloves
3. Safety goggles
4. Lubricants and gear box oil
5. Cleaning materials
6. Tube cleaning pressure washer
7. Self-sealing quick disconnect refrigerant hose fittings
8. Refrigerant recovery/recycling equipment
9. EPA/DOT approved refrigerant storage tanks
10. Dry nitrogen gas, cylinder, and regulator
11. Paint brushes
12. Approved refrigerant
13. Electronic Leak Detector
14. Megger
15. Variac.

TASKS

1. Lubricate drive couplings.
2. Lubricate motor bearings.
3. Lightly lubricate vane control linkage bearings, ball joints, and pivot points. DO NOT LUBRICATE the shaft of the vane operator.
4. Remove refrigerant in accordance with manufacturer's instructions. Use appropriate recovery equipment.
5. Drain and replace oil compressor oil reservoir including filters, strainers and traps.
6. Drain and replace purge compressor.
7. Drain and replace oil in gearbox. Check and clean oil strainer.

8. Check and correct alignment of drive couplings.
9. Inspect cooler and condenser tubes for scale. Clean if required.
10. Clean all water strainers in the system.
11. Use oil-dry nitrogen to test for leaks per manufacturer's instructions. If leaks cannot be stopped or corrected, report leak status to supervisor.
12. Pull vacuum on refrigeration machine in accordance with manufacturer's instructions. Add refrigerant as required per specifications.
13. Megger compressor and oil pump motors and record readings.
14. Check dashpot oil in main starter.
15. Tighten all starter, control panel, motor terminals, overloads, and oil heater leads, etc.
16. Check all contacts for wear, pitting, etc
17. Check calibrate overloads, record trip amps and trip times.
18. Check and calibrate safety controls.
19. Clean up the work area. Properly recycle or dispose of materials in accordance with envi ronmental regulations.

Sump Pump

Special Instructions:

1. Strainer cleaning requires removal of pump unit and should be handled as a repair.
2. Excessive sediment and debris, not removed by flushing the pit should be handled on a project basis, and not considered under this guide.
3. Review manufacturer's instructions.
4. If the material removed from the pump is hazardous, contract your environmental, health and safety department office for disposal instructions.

Tools & Materials:

1. Tool group C
2. Cleaning equipment and materials. Consult the MSDS for hazardous ingredients and proper PPE.
3. Lubricants. Consul the MSDS for hazardous ingredients and proper PPE.

TASKS

1. Flush pit and pump out.
2. Check bail, floats, rods, and switches. (Make sure float operates as designed.)
3. Clean and inspect motor (if not submersible) and perform necessary lubrication. On submersible pumps and motors, perform PM as suggested by the manufac turer. Repack pump if needed.
4. Inspect check valves.
5. Clean up work area and remove all debris.

Switchboard, Low Voltage

Special Instructions:

1. Schedule work and notify all operating personnel. The initial maintenance work on new equipment should be six months after installation.
2. Caution: This work requires a total board outage and safe removal of all possible sources of electricity. Review one-line diagrams to be sure that all circuits have been located. Identify the breakers necessary to remove all voltage sources including feedback. All incoming and outgoing circuits from this bus must be safely cleared, including any voltage transformers. Upon completion of checkpoints #1 and #2 below, de-energize and lockout the switchboard bus.
3. All protective devices mounted in the switchboard should be tested at this time, using appropriate PM guide cards.
4. These tests shall conform to the appropriate ASTM procedures and the values used as standards shall conform to the manufacturers and ANSI Standards Specifications. Tools and

Materials:

1. Tool group C
2. Cleaning tools and materials
3. Vacuum
4. Calibrated torque wrench
5. Insulation resistance test set.

TASKS

1. Perform a complete visual inspection. Look for:
   a. Proper alignment, anchorage and equipment grounding.
   b. Grounds or shorts.
   c. Evidence of overheating or arcing.
   d. Cable arrangements and supports cracked or damaged insulators.
2. Perform an infrared scan of the complete switchboard and all protective devices while it is energized.
   a. Remove cover plates.
   b. Inspect this during times of heaviest loading, if possible
   c. Record locations of hot spots. Note the temperature rise (Delta_T) of any found.

NOTE: Any connection with a Delta_T above 20 degrees should be corrected immediately.

3. Upon accomplishing special instruction #2 above, thoroughly vacuum all dust and dirt. Wipe clean the interior of the switchboard, including buses, insulators, and cables.
4. Inspect fuse clips for tightness and alignment.
5. Torque cable and bus connections to factory specifications, paying special intention to hot spots shown on the infrared scan. Hot circuits could be an indication of overloaded circuits or unbal anced loads.
6. Perform an insulation resistance test from phase to ground on each bus. Compare the results with previous tests to detect any weakening trend.
7. Refinish any damaged surfaces found.

Tank, Fuel Oil

Special Instructions:

1. If person must enter tank, test for oxygen deficiency, and supply proper respirator as needed.
2. Safety harness must be worn.
3. Review manufacturer's instructions.
4. Review the Standard Operating Procedures for "Confined Space Entry."
5. Review the Standard Operating Procedures for "Selection, Care, and Use of Respiratory Protection".

Tools and Materials:

1. Tool group C
2. Goggles
3. Respirator
4. Safety harness.

TASKS

1. Prior to end of heating season, adjust oil deliveries so oil will be nearly consumed.
2. Remove manhole.
3. Pump oil tank down within 6 inches of tank bottom.
4. Pump sludge from bottom of tank and flush. Dispose of appropriately. If material removed from the tank is hazardous waste, contact the Regional S and EM office for instructions.
5. Disconnect heating coil, remove from tank and clean.
6. Examine tank for leaks, and condition of piping connections.
7. Clean and adjust oil transfer pumps.
8. Examine, clean, and adjust operation of strainers, traps, control valves, oil flow meter, oil temper ature and pressure gauges.
9. Check floats and leveling devices in tank. Check float adjustment with depth level indicators.
10. Clean breather vents, conservation vents, and flame arrestors where appropriate.
11. Clean up work area and remove all debris.

Valve, Manually Operated

Tools and Materials:

1. Tool group C
2. Lubricants. Consult the MSDS for hazardous ingredients and proper PPE.

TASKS

1. Operate valve in full open/closed position. Loss of ability to close tightly will require inspec tion of valve seats and discs for wear and contaminant buildup.
2. Check for sticking valve stems and lubricate stems and fittings sparingly.
3. Replace packing; dress, re-bush, or replace packing gland assembly, if required.
4. Check for freedom of motion on valves equipped with wheel and chain for remote operation.
5. Clean up work site.

Valve, Motor Operated

Tools and Materials:
1. Tool group C
2. Lubricants. Consult the MSDS for hazardous ingredients and proper PPE.
3. Cleaning equipment and materials. Consult the MSDS for hazardous ingredients and proper PPE.

TASKS
1. Clean unit and make visual examination of all parts.
2. Operate from limit-to-limit. Observe operation; look for binding, sluggishness, action of limits, etc.
3. Determine if valve seats and holds properly.
4. Check condition of dials and positioners.
5. Check condition of packing.
6. Apply graphite to moving parts of valve.
7. Lubricate motor and gearbox as necessary.
8. Inspect contacts, brushes, motor, controls, switches, etc. Clean and adjust as necessary.
9. Clean up work site.

Valve, Regulating

Tools and Materials:
1. Tool group C
2. Cleaning equipment and materials. Consult the MSDS for hazardous ingredients and proper PPE. TASKS 1. Inspect for dirt collected at bleed port and restriction elbow. Clean if necessary.
2. Inspect joints for leakage. Tighten all bolts.
3. Check for dust or other material that may have sifted onto the upper face of the pilot pressure plate.
4. Remove and clean line strainer (back-flush where possible).
5. Inspect valve head and seats for nicks and abrasions. Notify supervisor if valve requires regrinding.
6. Inspect pressure reading against set point.
7. Check for free operation of valve stem
8. Inspect condition of diaphragm.
9. Inspect pilot line for leaks.
10. Clean up work site and remove all debris.

Valve, Safety Relief

Special Instructions:

1. Safety relief valves are designed to be operated by steam and should only be tested when sufficient pressure exists to clear the seating area of any debris.
2. Check with foreman and operating personnel before performing this test.

TASKS 1. Inspect condition of spring, flanges, and threaded connections.
2. Inspect and hand lift the manual lifting lever, checking for binding of the stem or seat.

Note that valve returns to proper position when lever is released.

3. Inspect support brackets and tighten as required.
4. Check that the discharge piping support is tight and not causing stress on the valve.
5. Clean the valve body.
6. Lubricate the stem and lever pivot.

Water Softener

Special instructions: 1. Review manufacturer's instructions.
2. Schedule service with operating personnel.
3. Secure and tag associated steam and water valves.
4. Allow the tank to cool before starting work.

Tools and Materials: 1. Tool group C
2. Grinding compound and lapping block
3. 12 volt drop light.

TASKS

1. All tanks.
   a. Drain the tank.
   b. Examine the exterior of tank, including fittings, gauges, manholes, and handholds for signs of leaks or corrosion. Correct as needed.
   c. Inspect structural supports and insulation or coverings for defects or deterioration.
   d. Open tank and remove rust or chemical deposits from interior tank surfaces.
   e. Remove and clean all spray nozzles.
   f. Thoroughly inspect interior of tank for pitting, cracking, and other defects.
2. Lime Water Softener.
   a. Dismantle vacuum breakers. Inspect stem, valve seat, and spring. Lap seat if required. Reassemble.
   b. Inspect, clean, and flush nozzle ring.
   c. Remove vent condenser heads and clean tubes.
   d. Inspect and clean sight glass, level indicators, and level controllers.
3. Zeolite Water Softener.
   a. Check filter bed for proper level.
   b. Take sample of zeolite resin according to manufacturer's instructions and send to lab for analysis.
   c. Check the operation of the multiport valve.
4. Anthracite water softener.
   a. Check the filter bed for proper level.

My wife loves to sew so I thought I'd add in a PM for her. Notice that good maintenance practice doesn't depend on the equipment. "Most sewing machine problems can be traced to poor general maintenance or neglect. To keep your machine in tiptop shape requires only a few simple supplies and a few minutes of attention daily, weekly, or monthly — depending on how much you sew." Sally Hickerson. That is pretty interesting advice since the same can be said for most industrial equipment.

Preventive Sewing-Machine Maintenance

From the pages of Threads Magazine

D	Keep it covered Exclude dirt	
4H	Change your needles often	Replace the needle after every four hours of sewing time
D	Wind bobbins correctly	Be sure there are no thread tails hanging from the bobbin when it's inserted into the bobbin case. These threads can jam the machine and cause the upper thread to break. And note that there's no such thing as a generic bobbin. Always use a bobbin designed for your machine
30H	Regular cleaning is essential	Start at the top and clean the tension disks with a folded piece of fine muslin. Be sure the presser foot is up, so that the tension springs are loose and the muslin can move easily between the disks, dislodging any lint or fuzz. Use a can of compressed air, blowing from back to front, to remove loose particles from around the tension disks and to clean other areas inside the machine. Don't blow into your machine yourself because breath contains moisture and will eventually cause corrosion. Get into the habit of removing the machine's needle and throwing it away after completing a project. Then take out the throat plate, bobbin, bobbin case, and hook race if this applies to your machine (new computerized machines do not have removable hooks). Clean under the feed dogs and around the bobbin area with a small brush, and use the compressed air to blow out any lint from inside of the bob bin case. The description continues…
30H	Lubricate	Use light oil recommended for sewing machines; do not use three-in-one oil. Check with your manual regarding any other areas on your machine that may require oiling, and use only a small drop for each spot. It is always better to oil too little more often than too much at one time. Avoid oiling any plastic parts.
2Y or 250H	PM routine (outsource)	I recommend a check-up by your dealer or an authorized mechanic every two years. Your machine will give you years of service if you take the time to care for it properly.

Sewing machines are used irregularly so the PM is based on hours of use. If you sew everyday you can translate the interval to a calendar.

Ice Machine PM, with thanks to Mitch Rens

General inspection	
W Clean around machine	Make sure that nothing (boxes, etc.) is stacked on or around the ice machine
W Insure free flow of Air	Make sure the machine is not at all covered during operation. There must be adequate airflow through and around the machine, to ensure long, competent life and maximum ice production.
W Check all water fittings and lines for leaks	Small leaks turn into large leaks
Exterior cleaning	
M Clean outside of unit	Sponge any dust and dirt off the outside of the machine with mild soap and water. Wipe it dry with a soft, clean cloth. Caution: Stainless steel panels should be cleaned with a mild soap or a commercial stainless steel cleanser. Remove heavy stains with stainless steel wool. Never use plain steel wool or abrasive pads, which will scratch the panels and cause rusting.
Cleaning the condenser	
Q* Lock Out Machine	Safety warning: Disconnect the electric power to the machine and the remote condenser at the electric service switch before cleaning the condenser.
Q* Clean the condenser at least self- every six months	The condenser fins are sharp, so use care when cleaning them. In contained and remote air-cooled models, a dirty condenser restricts air flow, resulting in excessively high operating temperatures. These higher temperatures reduce ice production and shorten component life.
Q* Clean the washable aluminum filter with mild soap-and-water	The washable aluminum filter on self-contained machines is designed to catch dust, dirt, lint, and grease and helps keep the condenser clean.
Q* Clean the outside of the condenser (the bottom side of the remote condenser)	Use a soft brush or vacuum with a brush attachment. Brush or wash the condenser from top to bottom, not from side to side. Be careful not to bend the fins. Shine a flashlight through the condenser to check for dirt between the fins. * All tasks to be performed monthly in a dirty environment
Q* Clean the condenser and water-regulating valve. (In water cooled units)	May require cleaning due to scale build up. *More often if water is hard, less often if water is soft

In every field we can see the push toward Preventive Maintenance being necessary for good quality output and long life. Ampex is one of the premier builders of reel-to-reel tape decks. They say: http://ampthetex.topcities.com/Tapemaintenance.htm

D	Cleaning of all the heads Record, Playback, and Erasure.	Clean all the other metal parts that come into contact with the tape. Use 100% pure alcohol or isopropyl alcohol. Symptoms of non-daily maintenance are a loss of the high frequency response, plus severe head and tape wear (uneven headwear).
W	Demagnetising	Do not bring a demagnetiser into contact with any metal parts on the tape machine. The machine must be switched off when demagnetising. I If demagnetising is to be done with the machine on, all the channels must be muted or the master fader must be turned right down, there is a risk of blowing up the speakers. Make sure that there is no storage media within 3ft of the demag netiser, to avoid the possibility of data corruption.
M	Alignment	Align the tape machine, using an alignment tape. .

Sample of a high technological PM for a device in a University Lab. In this PM

B-M is bi-monthly (every 2 months)
S-A is semi-annually (twice a year)
2A is every two years

Matrix 402 Etcher: Preventive Maintenance

M	If necessary, replace the extend- retract motor
M	Tighten the Hall effect switch clamps
M	Clean card reader
M	Calibrate the RF generator/ process controller board
B-M	Clean chamber
B-M	If necessary, replace the chamber pins
B-M	Clean RF gasket material
B-M	Refill the vacuum oil
B-M	Check the process and support gas regulators
B-M	Visual Orbitran check
B-M	Check vacuum integrity
B-M	Wipe down the main console assembly
B-M	Replace the lamps for the EMO, ON/OFF switch, etc.
B-M	Ensure cooling fans are operational
B-M	Check the coolant fluid level
Q	Pressure and gas flow check
Q	Clean the Orbitran gears
Q	Replace the extend- retract motor of the Orbitran
Q	Check and adjust the motion of the pins
Q	Matching network inspection
Q	Pick vacuum sensor check
Q	Check the Orbitran UP/DOWN limit stop
Q	Transport interface adjustment
Q	Adjust the door open/ close speed
S-A	Remove and clean the butterfly valve assembly
S-A	Remove and clean the vacuum isolation valve and replace the O-rings
S-A	Transport alignment procedure
S-A	Check the tightness of the chamber exhaust fitting
S-A	Clean the Clippard valves
S-A	If necessary, replace the Orbitran rotate motor
A	Calibrate the capacitance manometer
A	Replace the inline gas filter
A	Replace the thermocouple
A	Flush the water Recirculator and replace the coolant fluid
A	Replace the exhaust port O-ring 2
A	Service the Orbitran
2A	Check and adjust the DC voltages
2A	Remove, clean calibrate and adjust the MFC's
2A	Calibrate the DGH interface module

Appendix B: Glossary

Asset: A machine, building, or a system. An asset is the basic unit of maintenance. It could be a machine, piece of equipment, area (floor in a building), product production line, or even a major component.

Backlog: All work available to be done. Backlog work has been approved, parts are either listed or bought, and everything is ready to go.

Cause: (Special to FMECA) A cause is the means by which a particular element of the design or process results in a Failure Mode.

CM: See corrective maintenance

Capital spares: Usually large, expensive, long lead-time parts that are capitalized (not expensed) on the books and depreciated. These items are protection against downtime.

Call Back: Job where the maintenance person is called back because the asset broke again or the job wasn't finished the first time. See rework.

Charge-back: Maintenance work that is charged to the user. All work orders should be costed and billed back to the user's department. The maintenance budget is then included with the user budgets. Also calling rebilling.

Charge rate: The rate in dollars that you charge for a mechanic's time. In addition to the direct wages you add benefits and overhead (such as supervision, clerical support, shop tools, truck expenses, supplies). You might pay a tradesperson $15.00/hr and use a $35/hr or greater, charge rate.

Continuous Improvement (in maintenance): Reduction to the inputs (hours, materials, management time) to maintenance to provide a given level of maintenance service.

Core damage: Describes a normally rebuildable component that is damaged so badly that it cannot be repaired.

Corrective maintenance (CM): Maintenance activity that restores an asset to a preserved condition. Normally initiated as a result of a scheduled inspection. See planned work .

Criticality (Special to FMECA): The Criticality rating is the mathematical product of the Severity and Occurrence ratings. Criticality = (S) _ (O). This number is used to place priority on items that require additional quality planning.

Customer: Customers are internal and external departments, people, and processes that will be adversely affected by product failure.

Deferred maintenance: All the work you know needs to be done that you choose not to do. You put it off, usually in hope of retiring the asset or getting authorization to do a major job that will include the deferred items.

Detection (Special to FMECA): Detection is an assessment of the likelihood that the Current Controls (design and process) will detect the Cause of the Failure Mode or the Failure Mode itself, thus preventing it from reaching the Customer.

DIN work: ' Do It Now' is non-emergency work that you have to do now. An example would be moving furniture in the executive wing.

Effect Cause: (Special to FMECA) A Cause is the means by which a particular element of the design or process results in a Failure Mode. An Effect is an adverse consequence that the Customer might experience. The Customer could be the next operation, subsequent operations, or the end user.

Emergency work: Maintenance work requiring immediate response from the maintenance staff. Usually associated with some kind of danger, safety, damage, or major production problems.

Failure Mode: Failure Modes are sometimes described as categories of failure. A potential Failure Mode describes the way in which a product or process could fail to perform its desired function (design intent or performance requirements) as described by the needs, wants, and expectations of the internal and external Customers.

Feedback: (When used in the maintenance PM sense) Information from your individual failure history is accounted for in the task list. The list is increased in depth or frequency when failure history is high and decreased when it is low.

FMEA Element: FMEA elements are identified or analyzed in the FMEA process. Common examples are Functions, Failure Modes, Causes, Effects, Controls, and Actions. FMEA elements appear as column headings on the output form.
Frequency of Inspection: How often do you do the inspections? What criteria do you use to initiate the inspection? See PM clock.

Function: A Function could be any intended purpose of a product or process. FMEA or RCM functions are best described in verb-noun format with engineering specifications.

Future Benefit PM: PM task lists that are initiated by a breakdown rather then a normal schedule. The PM is done on a whole machine, assembly line, or process, after a section or sub-section breaks down. This method is popular with manufacturing cells where the individual machines are closely coupled. When one machine breaks, the whole cell is PM'ed.

Iatragenic: Failures that are caused by your service person.

Inspectors: The special crew or special role that has primary responsibility for PM's. Inspectors can be members of the maintenance department or can be members of any department (machine operators, drivers, security officers, custodians, etc.)

Inspection list: see task list

Interruptive (task): Any PM task that interrupts the normal operation of a machine, system or asset.

Labor: Physical effort a person has to expend to repair, inspect, or deal with a problem. Expressed in hours and can be divided by crafts or skills.

Life Cycle: This term denotes the stage in life of the asset. The author recognizes three stages: start-up, wealth, and breakdown.

MTBF: Mean time between failures. Important statistic to help set-up PM schedules and to determine reliability of a system.

MTTR: Mean time to repair. This calculation helps determine the cost of a typical failure. It also can be used to track skill level, training effectiveness, and effectiveness of maintenance improvements.

Management: The act of controlling or coping with any problem.

*Maintainability Improvement: Also Maintenance Improvement. Maintenance engineering activ-*ity that looks at the root cause of breakdowns and maintenance problems and designs a repair that prevents breakdowns in the future. Also includes improvements to make the equipment easier to maintain. .

Maintenance: The dictionary definition is "the act of holding or keeping in a preserved state." The dictionary doesn't say anything about repairs. It presumes that we are acting in such a way as to avoid the failure by preserving the asset.

Maintenance Improvement: Actions taken to reduce the amount of maintenance needed or actions taken to reduce the time for existing tasks.

Maintenance Prevention: Maintenance free designs resulting from increased effectiveness in the initial design of the equipment.

Non-interruptive task list: PM task list where all the tasks can safely be done without interrupting production or use of the machine.

Non-Scheduled work: Work that you didn't know about and plan for at least the day before. Falls into three categories: 1. Emergency 2. DIN 3. Routine Also work that you knew about but didn't think about in a systematic way and didn't add to a schedule.

Occurrence: Occurrence is an assessment of the likelihood that a particular cause will happen and result in the Failure Mode during the intended life and use of the product.

OEM: Original Equipment Manufacturer, the company that originally manufactured the equipment.

PCR: Planned Component Replacement. Maintenance authorizes component replacement on a schedule based on MTBF, downtime costs, and other factors. This technique fosters ultra-high reliability and is favored by the airline industry.

Parts: All the supplies, machine parts, and materials to repair an asset, or a system in or around an asset.

Planned maintenance: Maintenance work that has been reviewed and all resources and steps have been identified. Also see scheduled work

PM: Preventive Maintenance is a series of tasks that either, 1. Extend the life of an asset. 2. Detect that an asset has had critical wear and is going to fail or break down.

PM Clock: The parameter that initiates the PM task list for scheduling. Usually buildings and assets in regular use expressed in days (For example, PM every 90 days). Assets used irregularly may use other production measures such as pieces, machine hours, or cycles.

PM frequency: How often the PM task list will be done. The PM clock drives the frequency. See frequency of inspection.

PMO (PM optimization): Structured process to rationalize PMs so that all failure modes have PMs and there are no unnecessary PM tasks.

Predictive Maintenance: Maintenance techniques that inspect an asset to predict if a failure will occur. For example, an infrared survey might be done of an electrical distribution system looking for hot spots (where failures would be likely to occur). In industry, predictive maintenance is usually associated with advanced technology such as infrared measurements or vibration analysis.

Priority: The relative importance of the job. A safety problem would come before an energy improvement job.

Proactive: Action before a stimulus (Ant: reactive). A proactive maintenance department acts before a breakdown.

RCM: Reliability-centered Maintenance. A maintenance strategy designed to uncover the causes of low reliability and plan PM tasks to be directed specifically at those causes. RCM is a procedure for uncovering and overcoming failures.

Rework: All work that has to be done over. Rework is bad and indicates a problem in materials, skills, or scope of the original job. See call back.

Risk Priority Number: The Risk Priority Number is a mathematical product of the numerical Severity, Occurrence, and Detection ratings. RPN = (S) * (O) * (D). This number is used to place priority on items that require additional quality planning.

R*oot cause (and root cause analysis)*: The root cause is the underlying cause of a problem. For example you can snake out an old cast or galvanized sewer line every month and never be confident that it will stay open. The root cause is the hardened buildup inside the pipes, which necessitates pipe replacement. Analysis would study the slow drainage problem, figure out what was wrong, and estimate the cost of leaving it in place. Some problems (not usually this type of example) should not be fixed, and will be indicated by root cause analysis. .

Route maintenance: The mechanic has an established route through your facility to fix all the little problems reported. The route mechanic is usually very well equipped so most small problems.can be dealt with. Route maintenance and PM activity are sometimes combined.

Routine work: Work that is done on a routine basis where the work and material content is well known and understood. An example is daily line start-ups.

SWO: Standing Work Order, Work order for routine work. A standing work order will stay open for a week, month or more. The SWO for daily furnace inspection might stay open for a whole month.

Scheduled work: Work that is written-up by an inspector and known about at least 1 day in advance. The scheduler will put the work into the schedule to be done. Sometimes the inspector finds work that must be done immediately which becomes emergency or DIN. Same as planned maintenance.

Se*verity:* Severity is an assessment of how serious the Effect of the potential Failure Mode is on the Customer.

Short Repairs: Repairs that a PM or route person can do in less than 30 minutes with the tools and materials at hand. These are complete repairs and are distinct from temporary repairs.

String based PM: Usually simple PM tasks that are strung together on several machines. Examples of string PM's would include lubrication, filter change, or vibration inspection routes.

TPM: Total Productive Maintenance. A maintenance system set-up to eliminate all the barriers to production. TPM uses autonomous maintenance teams to carry out most maintenance activity.

Technical Library (Maintenance Technical Library): The repository of all maintenance information including (but only limited by your creativity and space) maintenance manuals, drawings, old notes on the asset, repair history, vendor catalogs, MSDS, PM information, engineering books, shop manuals, etc.

Task: One line on a task list (see below) that gives the inspector specific instruction to do one thing.

TLC: (Tighten, Lubercate, Clean) Basic good maintenance practice.

Task List: Directions to the inspector about what to look for during that inspection. Tasks could be inspect, clean, tighten, adjust, lubricate, replace, etc.

UM: User Maintenance. Any maintenance request primarily driven by a user. It includes breakdown, routine requests, and DIN jobs.

Unit: The asset that the task list is written for in a PM system. The unit can be a machine, a system, or even a component of a large machine.

Work Order: Written authorization to proceed with a repair or other activity to preserve a building.

Work request: Formal request to have work done. Can be filled out by the inspector during an inspection on a write-up form or by a maintenance user. Work requests are usually time/date stamped.

Appendix C: Resources

(this is not designed as an exhaustive list of resources but as a starting point for your own research)

http://www.edatamanage.com/companyinfo.htm Software for management of route maintenance

http://www.chevron.com/prodserv/nafl/intsol/content/lubeit.shtml#top
Chevron is one of the big vendors in the predictive and lubrication field.

http://www.bently.com/bnc/brochures/lube.htm
Bently LUBE™ Lubrication Data Management Software

http://www.lubecouncil.org/index.htm Email: info@lubecouncil.org
International Council for Machinery Lubrication
3728 South Elm Place, PMB 326, Broken Arrow, OK 74011-1803
Phone: (918) 451-7849 FAX: (918) 451-8139
Also http://www.lubecouncil.org/MLTI/mlt1cert.asp for their job descriptions

Kender (Group) info@kender.ie Automated Lubrication management equipment
Upper Mell, Drogheda, Co. Louth, Ireland Phone: 041-9838166 Fax: 041-9833754

RELCODE is software to aid in the analysis of equipment replacement. The vendor is Oliver-Group in Canada http://www.oliver-group.com/html/relcode.html.

Realty Times site. This is a newsletter for the property industry. This particular article gives good examples for PM in buildings http://realtytimes.com/rtnews/rtcpages/20020508_hoamaintenance.htm

PM Optimization http://www.pmoptimisation.com.au/default.shtml This is Steve Turner's site.

www.Plant-maintenance.com is a great resource down under and a Good group of articles from maintenance professionals from around the world:
http://www.plant-maintenance.com/maintenance_articles_rcm.shtml

For transfer switches and generator sets look at http://www.loftinequip.com/index.html for information

One of the most interesting sites is http://www.maintenance-tv.com/. This is a consultancy owned by ABB They maintain an interesting list of articles on common issues:
http://www.maintenance-tv.com/servlets/KSys/92/View.htm Their Flash Audit is very interesting and can provide useful information http://www.maintenance-tv.com/world/mtv/selfaudit/info.htm They also maintain a super site at: http://www.maintenance-tv.com/world/mtv/articles/articlesandlinks.htm

Additional information on FMECA can be found at http://www.fmeca.com/

Maintsmart CMMS Very savvy program for analysis of your failure data and turning it into information that can be used for PM design. http://www.maintsmart.com/ Phone: Toll-Free in the U.S. 1-888-398-0450, Outside the U.S. 1-209-367-0450, Fax: 1-209-369-9396, 216 South Fairmont Ave., Lodi, CA, 95240

www.maintenanceresources.com is one of the maintenance super sites with resources of all kinds for maintenance professionals interested in PM

http://www.reliabilityweb.com/index.htm Another excellent on-line resource for reliability

http://www.infrared-thermography.com/ They have a great site for infrared images. They are a full service infrared contractor with a national (USA) presence.

http://www.flirthermography.com/rentals/ Interested in used infrared equipment or rentals? Flir is one of the leading infrared camera manufacturers. We always recommend renting a prospective camera before buying it.

http://www.snellinfrared.com/ Training Company for Infrared

http://www.machinerylubrication.com This is Machinery Lubrication magazine.

http://www.practicingoilanalysis.com
Sign up here for a Newsletter with tips on lubrications http://www.oilanalysis.com/publications.asp

Insight Services contact: http://www.testoil.com/frame_freeoffer.html for free oil analysis test kit.

http://www.vib.com/ Vibration Specialty Corporation 100 Geiger Road Philadelphia PA 19115 USA Tel 215.698.0800 Fax 215.677.8874 Has WinProtect smart vibration analysis software.

Inuktun is a manufacturer of miniature cameras and camera transport systems and lights. http://www.inuktun.com/ Inuktun Services Ltd., 2569 Kenworth Road, Suite C, Nanaimo, BC Canada, V9T 3M4, Tel: (250) 729-8080

National Industrial Supply 1201 Rochester Road, Troy, MI 48083 http://www.nischain.com/ this is where I found a Magna-Flux PM for hooks. They are suppliers of chain and hooks.

Erik Concha web site http://home.earthlink.net/~eaconcha/Main_page_frame.htm offers a complete course in vibration analysis.

Trade group: Electrical Apparatus Service Association http://www.easa.com/ members service motors and other apparatus. Many members perform PdM such as sophisticated PdM

WWW.Emaint.com is a CMMS software company. Their product is available for LANS (company networks) and on the web as an ASP (Application Service Provider). They provided their PM library for this book.

Maintenance Technology Magazine can be found at www.mt-online.com. They are a good source for articles, research and vendor lists in predictive maintenance.

Society of Tribologists and Lubrication Engineers sponsors certifications in lubrication and oil analysis. They can be found at www.stle.com

The Vibration Institute has courses and certifications in all aspects of vibration analysis and can be found at www.vibinst.org

Technical Associates of Charlotte is an old-line engineering, training and consultant company. At www.technicalassociates.net they have complete offerings in vibration analysis, alignment, noise, and related (more technical) disciplines.

Academy of Infrared Thermography www.infraredtraining.net

Infrared Training Center www.infraredtraining.com

Snell Infrared www.snellinfrared.com

Index